Andre Henrique Marques Luiz
Mara Gabriela Novy Quadri

Emulsion, texture and microstructure of chicken mortadella

Andre Henrique Marques Luiz
Mara Gabriela Novy Quadri

Emulsion, texture and microstructure of chicken mortadella

Effect of Basic Chemical Composition and Ingredients on the Physical-Chemical Characteristics of Chicken Mortadella

Imprint

Any brand names and product names mentioned in this book are subject to trademark, brand or patent protection and are trademarks or registered trademarks of their respective holders. The use of brand names, product names, common names, trade names, product descriptions etc. even without a particular marking in this work is in no way to be construed to mean that such names may be regarded as unrestricted in respect of trademark and brand protection legislation and could thus be used by anyone.

Cover image: www.ingimage.com

This book is a translation from the original published under ISBN 978-3-330-99555-0.

Publisher:
Sciencia Scripts
is a trademark of
Dodo Books Indian Ocean Ltd. and OmniScriptum S.R.L publishing group

120 High Road, East Finchley, London, N2 9ED, United Kingdom
Str. Armeneasca 28/1, office 1, Chisinau MD-2012, Republic of Moldova, Europe
Managing Directors: Ieva Konstantinova, Victoria Ursu
info@omniscriptum.com

Printed at: see last page
ISBN: 978-620-3-32619-2

SUMMARY

The meat industry has been exhaustively searching for quality products, a characteristic mainly related to the balance of the composition. As a result, there has been growing interest in the development of chicken meat products due to their high nutritional value, availability, price and emulsification characteristics. The basic chemical components of emulsion formation (protein, fat and moisture) and ingredients were evaluated for quality in the formulation of chicken mortadella.

The aim of this study was to evaluate these components on the physico-chemical characteristics of chicken mortadella. The effects of protein (12-16%), fat (19-30%) and moisture (54-65%) levels on cooking losses, texture and microstructure of chicken mortadella were studied. Nine formulations were developed using a mixture design, and the variables protein, fat and moisture had a significant effect on the cooking losses and texture profile of the chicken mortadella ($p<0.05$). Higher cooking losses were observed in the region of lower stability, with values between 5.90%±0.67 and 9.78%±0.10; according to microscopy, the formation of larger fat globules was observed.

In the region of greatest stability, reduced cooking losses were observed with values between 2.50 per cent±0.29 and 4.28 per cent±0.55 for compositions with a high protein content and lower moisture content. The homogeneity and distribution of the fat globules were attributed to an adequate protein content to emulsify the amount of fat added, which resulted in greater emulsion stability. Different ingredients were added to a base formulation (protein: 12%, fat: 26.5% and moisture: 61.5%) and their effects evaluated. Part of the protein content of chicken breast and fat was replaced in the formulation by the following ingredients: mechanically separated chicken meat (10-40%), concentrated soya protein (1-4%), cassava starch (1-4) and sodium tripolyphosphate (0.5g/100g of dough). The addition of mechanically separated chicken meat (MSM) had an influence from 10% onwards, with higher quantities (up to 40%) not significantly altering cooking losses and the texture profile of the chicken mortadellas; an increase in pH was observed. The concentration of soya protein concentrate (3 to 4%) proved effective in reducing cooking losses and maintaining the firmness and chewiness of the product.

Cassava starch (FM) in concentrations between 1 and 4 per cent considerably reduced cooking losses, slightly altering the pH and texture parameters of chicken mortadella, while the addition of sodium tripolyphosphate reduced cooking losses. One proposed formulation was comparatively close to the reference, Seara brand commercial chicken mortadella (MFC), especially in terms of chewiness. This study has shown that understanding the contribution of the effects of each basic constituent, as well as the main ingredients used in the manufacture of chicken mortadella, is fundamental to developing better products and solving problems in the industry.

Keywords: chicken breast, emulsion stability, texture, microstructure.

SUMMARY

CHAPTER 1

INTRODUCTION

Market trends have shown that poultry meat consumption will continue to grow worldwide due to factors such as health, price, availability and the development of new products (BARBUT, 2012). Chicken meat has generated interest due to its high nutritional and economic values, as well as its emulsification characteristics (ZORBA and KURT, 2006). As a result, the meat industry has been constantly modifying formulations in the search for better products, with quality being mainly related to the balance of the composition (ZORBA, 2006; DOMÉNECH-ASENSI et al., 2013).

Mortadella is a product consumed all over the world and in many regions of Brazil it is a very important part of the diet (HORITA et al., 2011). It is defined by Brazilian legislation as an industrialised meat product obtained from an emulsion embedded in a specific wrapping and subjected to the appropriate heat treatment. The main physico-chemical characteristics are a minimum of 12% protein and maximums of 30% and 65% fat and moisture, respectively (BRASIL, 2000).

Typically, an emulsion forms droplet sizes between 0.1-100 l-im with the incorporation of an emulsifier at the oil-water interface. Because they differ from a classic emulsion, meat emulsions are considered to be emulsion-type products. Their properties are mainly governed by temperature, composition and droplet size distribution. However, industrial emulsification is usually controlled empirically (MCCLEMENTS, 2005; CALDERON et al., 2007).

A meat product is a multiphase system formed by the comminution of lean meat, fat, salt and other ingredients (GORDON and BARBUT, 1997). In a meat product, the use of meats from different species has different characteristics and emulsifying properties, and fat has a significant influence on the binding capacity of the emulsion (MORIN et al., 2004; ZORBA and KURT, 2006). In the emulsification process, the myofibrillar proteins, which are the main emulsifying agents in meat, act to form an interfacial protein film that surrounds the fat globules within the meat emulsion. The properties of the protein film around the fat globules, the gelling properties of the protein matrix and the physical characteristics of the fat contribute to fat stabilisation; on the other hand, water retention is influenced by pH and the amount of soluble protein. These proteins are extracted in the comminution process in the presence of salt, which favours their solubilisation by increasing the ionic strength of the medium. During heat treatment, fat globules and other elements are immobilised by the gelation of the protein matrix, favouring the appearance of a desirable texture, colour and flavour to the product (MANDIGO AND ESQUIVEL, 2004). Therefore, the texture will depend on the formation of this structure, which is influenced by the protein's ability to bind water, the presence of salt, the pH value, fat content and other ingredients (FLORES et al., 2007).

The above explains both the nutritional and economic importance of chicken meat, as well as the complexity of processing and the importance of formulating quality chicken mortadellas. Therefore, understanding the contribution of the effects of each basic constituent, as well as the main ingredients used in its manufacture, is fundamental to developing better products and solving problems in the industry.

1.1 OBJECTIVES

1.1.1 General Objective

The general aim of this study was to investigate the effect of chemical composition on the physico-chemical characteristics of chicken mortadellas.

1.1.2 Specific objectives

- to investigate the effect of the levels of the basic components protein, fat and moisture on the stability of the emulsion;
- to evaluate the effect of the addition of different ingredients on the physicochemical characteristics of chicken mortadella;
- to obtain information on the components responsible for the stability of the emulsion in the production of chicken mortadella;
- to develop a theoretical and experimental formulation that is comparable to commercial chicken mortadella.

CHAPTER 2

LITERATURE REVIEW

2.1 CHEMICAL COMPOSITION OF MEAT

Meat tissues are made up of five main chemical constituents: protein, lipids (fat), moisture (water), carbohydrates and inorganic matter (ash and minerals), with an average of approximately 19% protein, 2.5% lipids, 75% water, 1.5% non-protein nitrogen compounds and 1% inorganic matter. Factors such as animal species, breed, sex, type of feed and the muscle analysed, as shown in Table 1, influence the chemical composition of meat (KEETON and EDDY, 2004; ORDÓNEZ et al., 2005).

Table 1 Approximate chemical composition of some cuts of meat (%).

Animal	Cutting	Protein	Fat	Water	Ashes
SWINE	Palette	19,5	4,7	74,9	1,1
	Bacon	11,2	48,2	40,0	0,6
BOVINE	Thigh	21,8	0,7	76,4	1,2
	Loin	22,0	2,2	74,6	1,2
CHICKEN	Muscle	20,0	5,5	73,3	1,2
	Chest	23,3	1,2	74,4	1,1

Source: Adapted from ORDÓNEZ et al. (2005).

Even in small quantities, carbohydrates play an important role in post-mortem phenomena. Meat mainly contains glycogen and other carbohydrates in low concentrations. Thus, during the maturation process, glycogen is transformed into lactic acid, which can be monitored by measuring the pH. Meat also contains free amino acids, peptides, creatine, creatinine phosphate, creatinine, vitamins, nucleotides and nucleosides, as well as enzymes and mineral substances such as potassium, phosphorus, sodium, magnesium, zinc, iron and copper, among others (ORDÓNEZ et al., 2005; PARDI et al., 2007).

2.1.1 Proteins

Because of their physiological function, proteins are the most important components of live muscle and in meat they are the main source of protein in the human diet. Chemically they are complex molecules, and their properties and functions depend on the number and relative position of the amino acids, linked by peptide bonds. The primary structure of the protein is determined by the sequence of amino acids (PRICE and SCHWEIGERT, 1971; PINCUS, 2001; KEETON and EDDY, 2004).

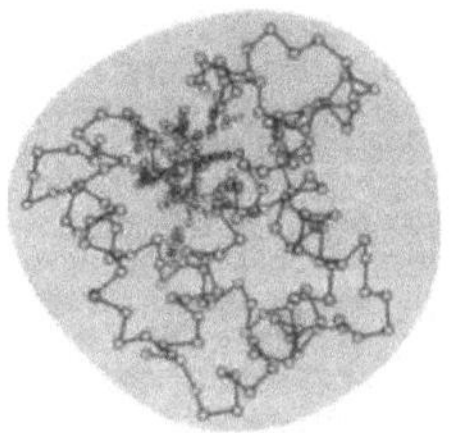

Figure 1 Representation of the globular structure of myoglobin.

Source: Adapted from TORNBERG (2005).

Generally speaking, the main proteins are classified as sarcoplasmic (metabolic), connective tissue proteins (support) and myofibrillar (contractile). The sarcoplasmic, or water-soluble, proteins are found in the sarcoplasm, a fluid that surrounds the myofibrils, composed predominantly of oxidative enzymes. The colour of muscle tissue depends on the concentration of myoglobin (Mb, 16 kDa) and the oxidation state of the iron atom: Fe^{2+}, myoglobin-H2O (purple red); $Fe^{(2+)}$, oxymyoglobin-O2 (bright red); $Fe^{(3+)}$, metmyoglobin-H2O (brown). Chicken and beef breast muscles have myoglobin concentrations of between 1 mg/g and 10 mg/g of muscle, respectively. Figure 1 shows a representation of the globular structure of myoglobin. Connective tissue proteins consist of a viscous solution of soluble glycoproteins with extracellular collagen and elastin fibres incorporated into the glycoprotein matrix. Myofibrillar proteins, or those soluble in saline solution, account around 60 per cent of the muscle's total proteins (Figure 2). Approximately 22% of the myofibrillar protein material is made up of actin (42 kDa) and 43% of myosin (520 kDa; 2 subunits of 220 kDa, 4 of 20 kDa) or 65% of the total amount of myofibrillar proteins. The isoelectric points of actin and myosin are 4.7 and 5.4, respectively. During rigor mortis, actin and myosin combine to form actomyosin and the muscle becomes rigid and loses elasticity. Also classified as myofibrillar proteins are tropomyosins (68 kDa), with 80-100% alpha helix, troponins, among others. Myofibrillar proteins are the most important for the emulsification process, influencing the water retention capacity, intermolecular bonding in gelation and mechanical stability of the emulsion (PRICE and SCHWEIGERT, 1971; HONIKEL, 2004; KEETON and EDDY, 2004; ORDÓNEZ et al., 2005; PARDI et al., 2007).

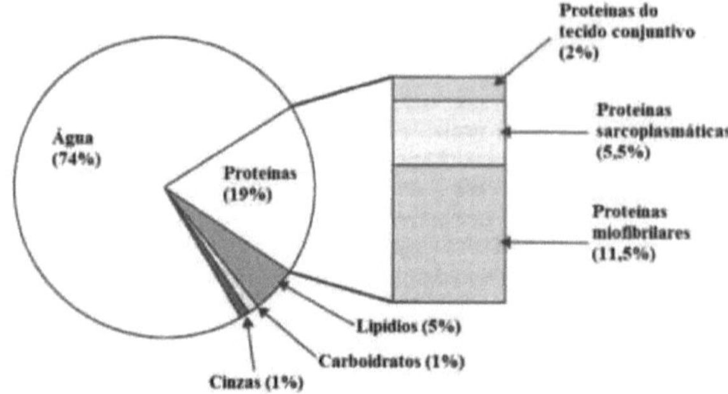

Figure 2 Composition of skeletal muscle tissue.

Source: Adapted from KEETON and EDDY (2004).

The structural unit of muscle is the muscle fibre. Skeletal striated muscle is made up of fibres, multinucleated cells, which form bundles surrounded by a connective membrane called the epimysium. This surrounds small bundles of fibres called the perimysium, from which septa lining each fibre called the endomysium. The muscle fibre is located within the sarcolemma, the membrane beneath the endomysium. Figures 3 and 4 show structural details of the muscle and muscle fibre. The myofibrils are long, slender cylindrical structures with a diameter of between 1 and 2 |im and are located within the sarcoplasm, with the nuclei distributed peripherally. The mitochondria, which are responsible for producing energy for the cell, are located between the myofibrils at the nuclear poles (PRICE and SCHWEIGERT, 1971; PARDI et al., 2007).

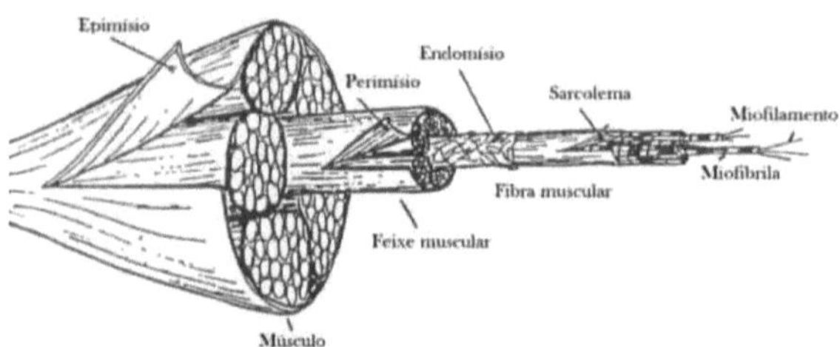

Figure 3 Schematic representation of a muscular structure.

Source: Adapted from PARDI et al. (2007).

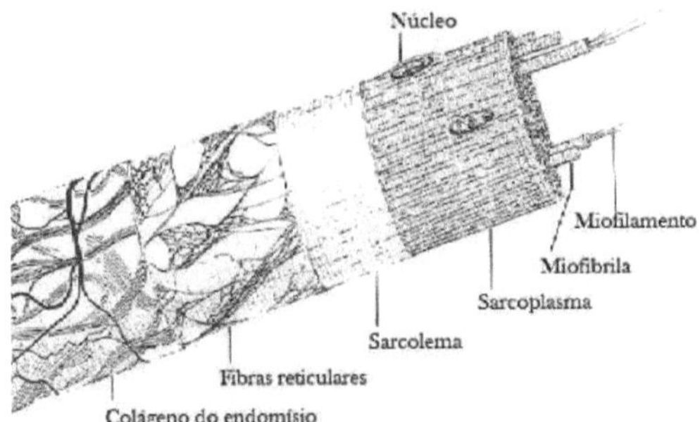

Figure 4 Schematic diagram of a muscle fibre.

Source: Adapted from PRICE and SCHWEIGERT (1971).

Myosin contains around 200-400 molecules in each of its thick filaments in a complex formation, with a displacement between molecules of 70-75nm, where the end of each molecule that has a globular head is directed towards the other end or another filament, leaving the centre of this region less dense. Its molecule contains between 60-70% alpha helix. The two subunits with the highest molecular weight are probably coiled like a rope of threads forming a 900-1200nm long bar that ends with a globular portion. The subunits with lower molecular weights appear to be located in the globular head of the molecule, shown in Figure 5. The normal muscles of vertebrate species (tuna, chicken, rabbit) have myosin molecules with similar enzymatic characteristics and amino acid composition. Myosin is easily soluble at ionic strengths greater than 0.3 and is sensitive to heat (PRICE and SCHWEIGERT, 1971; PARDI et al., 2007).

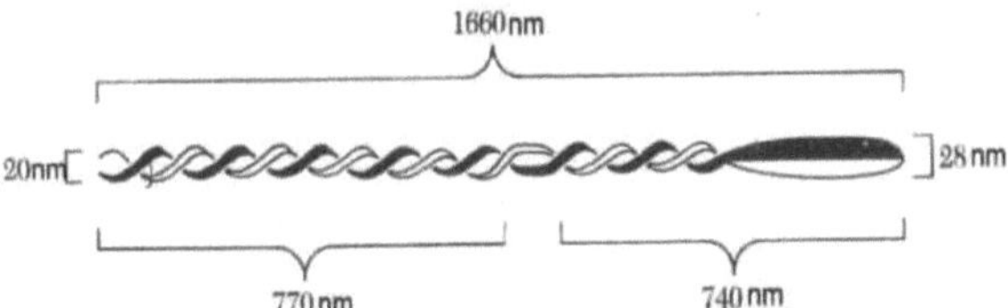

Figure 5 Schematic representation of the myosin molecule.

Source: Adapted from PRICE and SCHWEIGERT (1971).

1.1.2 Lipids/Fat

It is the most variable component of meat, both quantitatively and qualitatively. Fats are fundamentally made up of esters formed by glycerol and carboxylic acids. Saturated and monounsaturated fatty acids are the majority in triglycerides and vary mainly according to species and diet. Poultry fat is more unsaturated than that of pigs, cattle and sheep, according to Table 2 (PRICE and SCHWEIGERT, 1971; KEETON and EDDY, 2004; ORDOÑEZ et al., 2005). As well as contributing to palatability, fat also contributes to the structural stability of emulsified products (HEDRICK et al., 1994).

Table 2 Degree of saturation of the fatty acid components of muscle tissue lipids in different species.

Species	% Saturated	% Monoenoic	% Polyenoids
Cattle	40-71	41-53	0-6

Pigs	39-49	43-70	3-18
Sheep	46-64	36-47	3-5
Birds	28-33	39-51	14-23

Source: Adapted from ORDÓNEZ et al. (2005).

2.1.3 Humidity (Water)

In post-mortem muscle tissue, water is the main component, comprising 75-80% of the cell mass. Water significantly influences meat quality, mainly affecting roughness and tenderness, as well as changes that occur in the meat during chilling, storage and processing. Water is mainly present in lean muscle tissue and many of the physical properties depend on the water retention capacity, which is related to its final pH. The ionic environment and the degree of contraction of myofibrillar proteins are the main factors affecting the retention and loss of moisture in muscle tissue (KEETON and EDDY, 2004; ORDÓNEZ et al., 2005).

2.2 FUNCTIONAL PROPERTIES OF PROTEINS

The functional properties of proteins include texture formation and related properties such as: water retention, solubility/extraction, gelling, emulsification and adhesion/binding. Sarcoplasmic proteins have high surface activity and foam-forming capacity. Connective tissue proteins, basically collagen molecules, have a remarkable water-binding capacity and can play a role in shaping the texture of emulsified products. Myofibrillar proteins, on the other hand, have various functionalities: gelling, emulsification, water binding and adhesion. However, variations in quality attributes between muscle foods are mainly caused by the use of different ingredients and processing conditions. The amino acid composition, spatial arrangement of the protein structure, solubility characteristics, concentration of extracted proteins in the aqueous phase and hydrophobicity are some of the important intrinsic factors that can influence the functionality of proteins. On the other hand, extrinsic factors such as pH, temperature, cooking rate, degree of comminution, oxidation and the presence of other components, as well as other processing factors can significantly influence the functionality of proteins (XIONG, 2004).

The functionality of proteins in the process of emulsification, gelation and water binding is determined by molecular interactions: disulphide bridges, hydrogen bonds, electrostatic attractions and hydrophobic interactions. Disulphide bridges are important, for example, for the development of an acceptable texture, influencing the elasticity, cohesiveness and hardness of the product. Hydrophobic interactions, on the other hand, are essential for the formation of a cohesive and flexible gelling structure, as well as being crucial for the formation and stabilisation of the interfacial film of protein in the protein matrix. Hydrogen bonds and electrostatic attractions appear to play a role in binding fat globules to the interfacial protein film (MANDIGO and ESQUIVEL, 2004).

Proteins are compounds of particular interest due to their amphiphilic nature and film-forming capacity. While molecules with a lower molecular weight diffuse quickly to the interface, proteins, due to their bulkier nature and higher molecular weight, diffuse more slowly and, once at the interface, can develop strong viscoelastic films favouring the stabilisation of the emulsion (MCCLEMENTS, 2005; LAM and NICKERSON, 2013).

2.2.1 Water retention

Proteins play a crucial role in immobilising water in meat and meat products, contributing to the juiciness and tenderness of cooked meat. From a physico-chemical point of view, it can be in a free state or in a bound form, i.e. strongly associated with proteins through hydrogen bonds. In comminuted products, a large portion of water is also trapped in the gel matrix formed by the myofibrillar proteins. The use of salts (NaCl or KCl) and phosphates increases the interfilamentary spaces and expands the myofibrils, improving the water retention of meat and meat products. Different phosphates can be used alone or in combinations at a level of no more than 0.5 per cent of the final product weight. This process results in the weakening of the myofibrillar structure, thus allowing water molecules to diffuse more easily into the interfilamentary spaces, increasing the degree of hydration. Significant muscle fibre swelling is achieved when the concentration of NaCl is increased to approximately 0.6 mol/L, or 0.4 mol/L in the presence of phosphate. The degree of hydration continues until 1.0 mol/L (approximately 4%) is incorporated, at which point the fibre will begin to shrink (Figure 6) (XIONG, 2004).

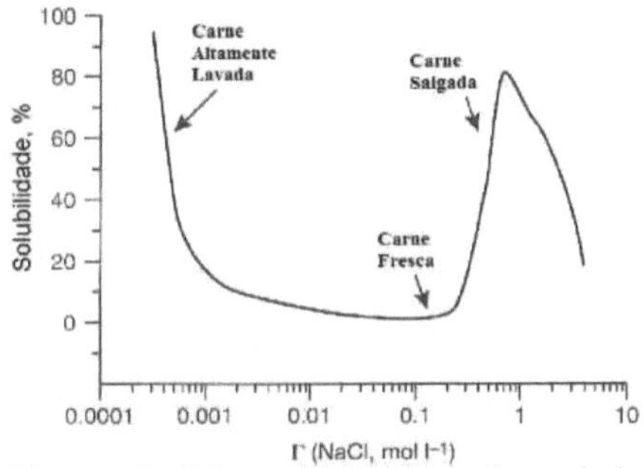

Figure 6 Schematic representation of the ionic strength: solubility relationship of muscle myofibrillar proteins.

Source: Adapted from XIONG (2004).

2.2.2 Gelation

In meat processing, the ability to form gel is one of the most important characteristics. It results from the unfolding and subsequent association of soluble protein molecules to form aggregates and filaments. When aggregation reaches a certain critical level, a gel with a three-dimensional structure is created and consists of cross-linked peptides with a large amount of trapped water. The gels formed act to stabilise the emulsified fat globules and trap other ingredients. The mechanism, under processing conditions (0.5-0.6mol/L NaCl, pH 6.0-6.5), has been extensively researched. As shown in Figure 7, the extracted myosin is utilised

as a model protein, and the sequential changes were induced by heat. At 35°C there is an unfolding of the myosin head with the formation of dimers and oligomers through head-head interaction. As the temperature rises to 40°C, a globular mass is formed. At 45°C, oligomers coexist with aggregates formed by the coalescence of two or more oligomers. Between 50 and 60°C, these oligomers aggregate further, apparently involving a tail-to-tail reticulation, to form particles that make up the filaments of the gel networks. Since the ordered aggregation of soluble proteins is continuous, processing conditions and formulations must be carefully controlled. The strength of the myofibrillar protein gel increases exponentially with concentration. In this sense, a slow heating process is desirable to allow progressive denaturation, favouring an ordered protein-protein association. The properties of the gels formed by myofibrillar proteins depend on the type of fibre. White fibre proteins have a greater tendency to gel on heating compared to red fibre proteins. This difference is attributed to structural issues and the solubility of different protein isomers, especially myosin, and can be explained by the formation of more elongated filaments by white fibre myosin compared to red fibres (XIONG, 2004).

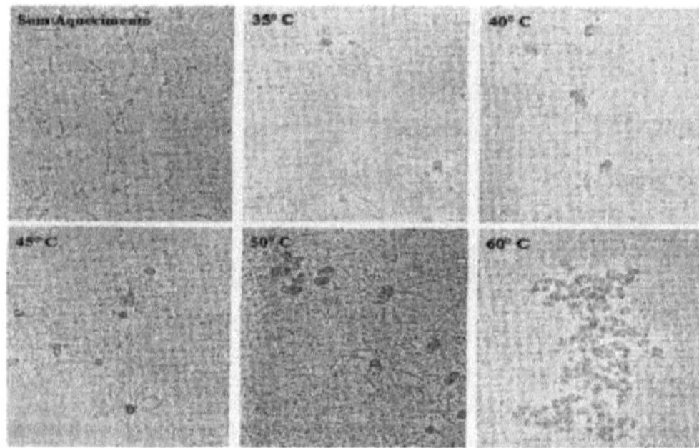

Figure 7 Micrographs of myosin aggregation during heat-induced gelation in 0.6 mol^{-1} KCL, pH 6.0-6.5.

Source: Adapted from XIONG (2004).

2.2.3 Emulsification

Typically, an emulsion is formed by two immiscible liquid phases where emulsification consists of the dispersion of one fluid in another by creating an interface. The properties of emulsions are mainly governed by temperature, composition and droplet size distribution. However, industrial emulsification is usually controlled empirically (MCCLEMENTS, 2005; CALDERON et al., 2007).

In a meat emulsion, fat globules are dispersed and stabilised in an aqueous matrix made up of salt-soluble myofibrillar proteins, muscle fibre segments and various other ingredients, as shown in Figure 8. Because they differ from a classic emulsion, they are considered to be emulsion-type products and are usually referred to as "meat batter", as they represent a multiphase system. During the comminution process, there is a structural orientation at the fat-water interface which is thermodynamically favourable due to the decrease in the total free energy of the system. With the reduction of particles and the extraction of muscle proteins, the formation of a flexible, viscoelastic membrane around the fat globules is essential for the stability of the emulsion. During emulsification, myosin and actomyosin are preferentially adsorbed at the fat-water interface. In such a system, the proteins are present in three distinct phases: the protein matrix, the aqueous phase and the interfacial protein film and the types and quantities of proteins present in each phase influence the textural properties and stability of the baked products. Myosin's superior emulsifying capacity is attributed to its structure showing an uneven distribution of polar and apolar amino acids in different segments of its molecule, a prevalence of hydrophobic residues in the head region and a preponderance of hydrophilic groups in the tail part, as well as having a high length to diameter ratio of approximately (40:1), a structure favourable to protein-protein interaction and molecular flexibility at the interface. Another mechanism for stabilising meat emulsions is the physical trapping of fat particles in the protein matrix, formed largely by protein-protein interaction. This means that the formation of a meat emulsion is due to a combination of an effective emulsifier film around the fat globules and the physical entrapment and bonding provided by the ordered network of proteins (MANDIGO and ESQUIVEL, 2004; XIONG, 2004).

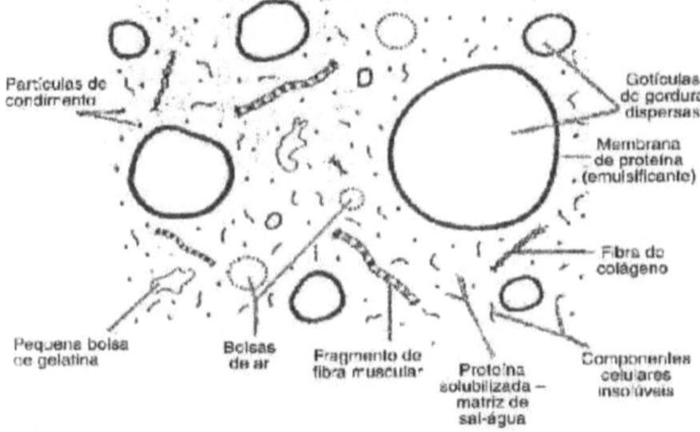

Figure 8 Schematic representation of a meat emulsion.
Source: LEMOS et al. (2011).

The formation and property of the interfacial protein film plays an important role in stabilising the fat. The protein gel formed also has high water-binding capacity and elastic properties. The protein-protein interaction is important for the formation of the soluble protein network and the hydrophobicity of the protein and high ionic strengths contribute to the formation of the interfacial film that encapsulates the fat globules. The right amount of emulsifying agent and smaller fat particle sizes favour emulsion stability. Short cutter times result in thick protein coatings around the fat globules. A better distribution of fat and protein along the interface favours stability. As the fat expands during heat treatment, if the protein film formed is too thin, the mechanical strength of the system decreases, which can lead to instability. Conversely, excessive amounts of protein in the continuous phase can decrease the flexibility of the product, as they favour the reduction of protein-protein interactions and proper gelation during cooking. Sufficient amounts of protein and fat seem to play an important role in emulsion stability. Products with reduced fat, on the other hand, require an increase in non-meat ingredients to improve functionality, especially for additional water binding. To summarise, emulsion stability is related to the thickness of the interfacial protein film, fat content and the integrity and density of the surrounding medium and its response to heat treatment. The type and characteristics of the fat also interfere with emulsion stability. Considering similar chain lengths, saturated fats are more easily emulsified, which may be related to their melting point, degree of dispersion and absorption of the fat by the interfacial protein film at the fat-water interface. The viscosity of the meat mixture decreases when the temperature is above the melting point of the fat, where the lower density of the fat particles favours surface displacement. In the face of thermal transitions, the protein matrix goes through various transition phases. Between 40-50°C, the process of myosin denaturation and gelation begins. At around 60°C, with a certain stability in the system, the collagen fuses. The structure is reinforced between 72 and 83°C, when actin is denatured. The fat within the system begins its melting process before the gelation of the protein matrix has been finalised. The presence of a sufficient emulsifier is therefore essential for stabilising the fat. Both the interfacial film and the structure of the protein matrix must be stable and cohesive, so that the expansion of the liquefied fat does not coalesce into pockets. In this case, what is known as emulsion breakage can occur as a result of the instability of the system, which can be quantified by the release of fat and/or water from the matrix during cooking. However, there are unlikely to be any visible signs before cooking indicating problems in the final product. Measuring the electrical conductivity of the raw dough gives an idea of the stability of the emulsion (MANDIGO and ESQUIVEL, 2004).

The emulsion is stabilised after cooking when fat globules and other components are immobilised by the gelation of the protein matrix and the characteristic properties of the final product are achieved. The stability of the emulsion can be measured by quantifying the loss of fat and moisture from the raw dough during cooking, due to the formation of channels through the matrix allowing this fluid to migrate to the surface of the product. Physical factors such as mixing, mixing time, cutter speed and external temperature influence the stability of the meat emulsion, as shown in Figure 9 (IGNÁCIO, 2011).

Figure 9 Schematic diagram of the main physical factors affecting the stability of meat patties. Source: Adapted from IGNÁCIO (2011).

2.3 MORTADELA

Mortadella is a product consumed all over the world and in many regions of Brazil it is a very important part of the diet (HORITA et al., 2011). It is defined by Brazilian legislation as an industrialised meat product obtained from an emulsion embedded in a specific wrapping and subjected to the appropriate heat treatment. The main physico-chemical characteristics are a minimum of 12% protein and maximums of 30% and 65% fat and moisture, respectively (BRASIL, 2000).

2.3.1 Legal aspects

According to the Brazilian technical regulation on identity and quality, mortadella is defined as an industrialised meat product produced from an emulsion of meat from butchered animals, with or without bacon, added ingredients, embedded in natural or artificial wrapping, in different forms and subjected to appropriate heat treatment. It is classified according to the composition of the raw material and manufacturing techniques. Specifically, poultry meat mortadella may contain in its composition: poultry meat, mechanically separated meat (max. 40%), fat and up to 5% of edible offal (liver, gizzard and heart). Mandatory ingredients in chicken mortadella are chicken meat and salt. Optional ingredients include water, animal and/or vegetable fat, vegetable (max 4%) and/or animal protein, intentional additives, binding agents, sugars, flavours, spices and seasonings, as well as vegetables (almonds, pistachios, fruit, olives, etc) and cheeses. Sensory requirements such as texture, colour, taste and smell are characteristic. The physico-chemical requirements are shown in Table 3. For additives and processing aids, contaminants, hygiene, weights and measures, labelling, physico-chemical analysis methods and sampling, the current regulations and criteria set out in this document apply (BRASIL, 2000).

Table 3 Physico-chemical characteristics for poultry mortadella

Features	
Total carbohydrates	max. 10%
Starch	max. 5%
Humidity	max. 65%
Fat	max. 30%
Protein	min. 12%
Calcium content Dry Base	max. 0.6%

Source: Adapted from BRASIL (2000).

2.3.2 Ingredients and Additives

Mechanically separated meat

Mechanically separated meat is obtained by the mechanical process of grinding and separating the bones of butchered animals and is used to make specific meat products. It has good emulsification capacity, emulsion stability and water retention capacity. It can be used in varying proportions according to current legislation (BRASIL, 2000; JIMÉNEZ COLMENERO, 2004).

Soya protein

Soya protein can be used as an ingredient in mortadella for economic, compositional (nutritional) and functional purposes, helping to form and stabilise the emulsion. Its incorporation influences the processing and physical-chemical properties of emulsified products. In this sense, soya is one of the most widely used proteins in meat products. Protein concentrates and isolates have good water and fat binding and emulsifying properties. The soya protein polymer chain contains both lipophilic and hydrophilic groups, promoting the formation of stable oil-in-water emulsions. Its action is also related to the protein-water interaction, which increases the viscosity of the dough by creating a gel matrix during heating. It can associate with various types of compounds, adhering to solid particles and acting as a suspending agent in solution. Under normal processing conditions (65-73°C; pH 5.5-6.0; ionic strength 0.1-0.6) none of the main soya globulins show any significant structural changes with muscle proteins. In high concentrations it can act as a diluent, weakening the gel-forming capacity of meat proteins, negatively affecting the texture of the final product, as well as imparting residual flavours, depending on the type of derivative used (JIMÉNEZ COLMENERO, 2004).Soya protein has polypeptide chains associated in a three-dimensional structure by disulphide and hydrogen bonds with a molecular weight of between 300-600 kDa. They can be divided into albumins, which are soluble in water, and globulins, which are soluble in saline solution, representing the majority of total proteins. Based on sedimentation coefficients, soya protein contains four main components: 2, 7, 11 and 15. In general, soya protein contains between 30-50% 11S, referring to glycinin (200-400 kDa), (Figure 10) and 20-30% 7S, conglycinin (100-200kDa), presenting similar secondary structures (SUN, 2005; NISHINARI et al., 2014).

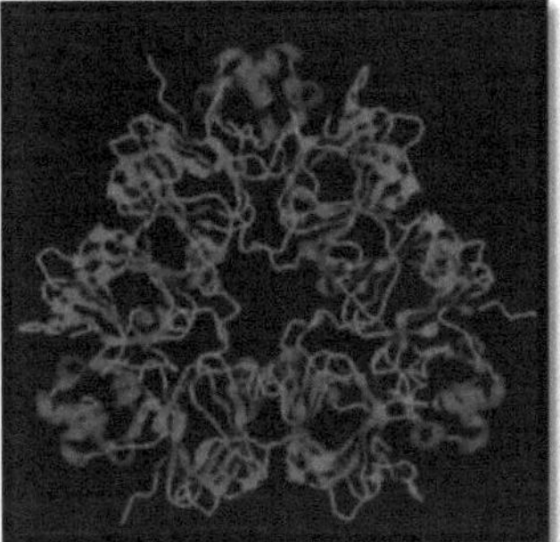

Figure 10 Ribbon representation: glycinin structure.

Source: SUN (2005).

Starch - Native structure

Starch is an ingredient with a high water-binding capacity and its source is mainly determined by the predominant crop in each region. Each granule is organised in a "ring" pattern of glucose chains: amylose and amylopectin polymers. Amylose (100-1000 kDa) is a low molecular weight linear polymer (Figure 11a). Amylopectin is a branched polymer with a high molecular weight ranging from 10000-1000000 kDa, as shown in Figure 11b. During cooking, water is strongly bound to the starch base and there is also a huge impact on the texture of the product. Due to the lower amount of amylose, cassava starch has a soft gel and is mainly used in extremely soft systems. Its gelatinisation temperature is between 62 and 73°C. There is no interaction between the starch and the meat protein and the most important criterion for choosing a starch is its gelatinisation temperature, which must correspond to the temperatures reached by the starch.

during the thermal process. In emulsified meat products such as mortadella, starches are useful when the meat is of lower quality or when high levels of water are used. Native starches have the disadvantage of being unstable during product storage. In general, a typical dosage of 2-3% of the final product is used, but higher concentrations are common for native starches (TESTER et al., 2004; JOLY and ANDERSTEIN, 2009).

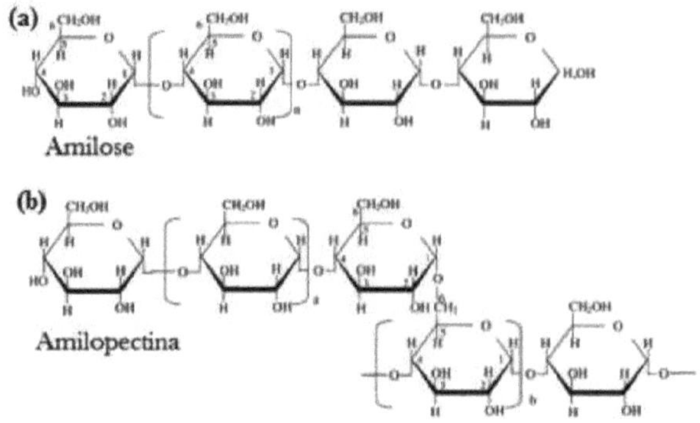

(a)

Amilose

(b)

Amilopectina

Figure 11 Structure of amylose and amylopectin.
Source: Adapted from TESTER et al. (2004).

Salt

Salt (sodium chloride) is widely recognised as a multifunctional ingredient in meat products and is mainly used to impart flavour with microbial inhibition, extend shelf life and solubilise proteins myofibrillar. The definition of the salt concentration in the final product is based on customer preferences, but values between 1.5 and 2.5 per cent are common in processed meat products. Microbial inhibition and a longer shelf life are achieved by reducing water activity. Increased water retention by the protein structure in the presence of chloride ion affects cooking yield, juiciness and tenderness of the product

consumed. It is much more important than the sodium ion in increasing water binding by meat proteins, as well as accelerating colour formation in cured products by increasing the rate of reaction of nitrite to nitric oxide. Water activity is significantly reduced with the addition of salt, favouring microbial inhibition and a longer shelf life. The sodium ion is responsible for the characteristic taste and for intensifying flavours (PEARSON and GILLETT, 1996; MILLS, 2004; SEBRANEK, 2009)

Phosphates

The phosphates used in the processing of meat products are generally alkaline polyphosphates and sodium tripolyphosphate is a linear polymer that is widely used industrially. Their action increases the pH of the medium, increasing the tendency for water to bind. They also act as metal ion chelators that favour lipid oxidation, as well as increasing the stability of emulsions and contributing to the organoleptic characteristics of products (MILLS, 2004; LEMOS et al., 2011).

Erythorbate

Erythorbate (sodium or potassium) is a reducing agent used in cured products to facilitate the reduction of nitrite into nitric oxide for the stabilisation of myoglobin.

High residual nitrite concentrations must be reduced to avoid the formation of nitrosamines, which are carcinogenic compounds formed by decomposition reactions (PEARSON and GILLETT, 1996; MILLS, 2004).

Spices

Spices are aromatic substances derived from vegetative plants or herbs that are not processed beyond drying, cleaning, sorting and grinding. The essential oils are protected in the cell structures and their content gives them a longer shelf life. The main spices used in meat emulsions are onion, garlic, nutmeg, black and white peppercorns, paprika, cinnamon powder and mace. Garlic (Allium sativum L.) belongs to the Liliaceae family and is commonly used in dried or powdered form with a penetrating odour and extremely pungent taste. Onion (Allium cepa) also belongs to the liliaceae family and is usually used together with garlic, forming the well-known home-made flavour (PEARSON and GILLETT, 1996; PARDI et al., 2007).

Nitrite

Nitrite is an essential additive in cured meat and can be supplied in the form of sodium nitrite (NaNO2) or potassium nitrite (KNO$_2$). It is highly reactive and functions as an oxidising, reducing or nitrosating agent. It is mainly used to develop the typical colour and flavour of cured meat, as an antioxidant, to prevent a rancid taste and as a bacteriostatic, acting synergistically with salt to prevent botulism (OCKERMAN and BASU, 2004; SEBRANEK, 2009).

2.3.3 Processing

The processing of mortadella follows a continuous sequence of events and each step is important for a successful operation. Grinding transforms pieces of meat of varying size, shape and fat content into uniform cylinders. The feed screw conveys the meat by pressing it into holes in the grinder plate, where the rotating blade cuts it, helping the meat to pass through the holes in the grinder disc. A mixer can then be used to favour the even distribution of lean meat and fat. Moving on to the cutter (Figure 12), which consists of a rotating container that holds the mass of meat, while rotating knife blades on a cutting shaft carry out its comminution. It is recommended that the final temperature of the mass does not exceed 15°C. In many cases, an emulsifier is used, combining the principles of grinding and cutting. Afterwards, the product is stuffed by extruding the emulsified meat mass into specific wrappings in order to shape the product. The inlaid casing is then tied up with wire or metal clips. Thermal cooking, usually carried out in ovens, where this process results in the coagulation of the proteins and the subsequent appearance of the product's texture (65 - 70°C). The cooking time must be long enough for the product to reach 72°C inside, i.e. for the product to be pasteurised. The product is then rapidly cooled in showers or cold water baths (0 to 4°C), followed by refrigerated storage (PEARSON and GILLETT, 1996; KNIPE, 2004; ORDÓNEZ et al., 2005).

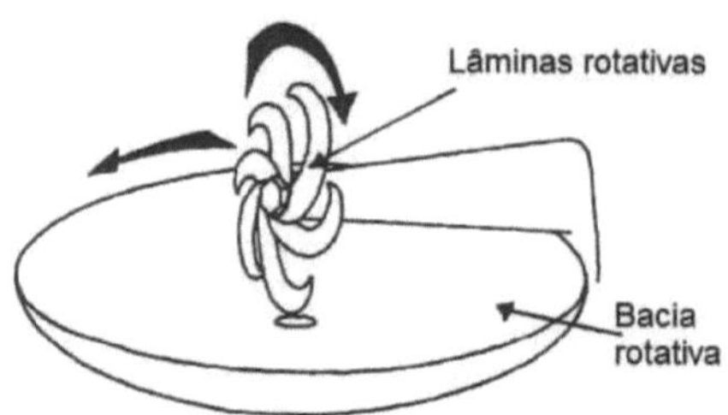

Figure 12 Schematic representation of a cutter.
Source: PEARSON and GILLETT (1996).

2.4 TEXTURE

Texture influences the overall acceptability of food and meat is classified in a critical group where texture quality is the dominant characteristic. Texture properties are considered to be physical characteristics dependent on the structure of the food, and are detected by the sensation of touch (deformation, disintegration, flow of the food under a force) and can be measured objectively with functions of mass, time and distance (BOURNE, 2002a).

The principle of texture profile analysis (TPA) is based on placing a sample of standard size and shape where it is compressed and decompressed twice by a cylinder. Figure 13 shows a typical texture profile curve generated by the texturometer. In this way, it was defined that: hardness is the height of the peak force over the first compression cycle (1^a bite). Fracturability is the significant breaking force in the curve at the first bite; Cohesiveness is the ratio between the areas of positive force between the first and second compression (A2/A1). Adhesiveness is the area of the negative force of the first bite (A3) which represents the work required to pull the plunger compressed. Elasticity is defined as the distance over which the food has regained its peak during the time between the end of the first bite and the end of the second bite. Gumminess is the product of hardness and cohesiveness. Chewiness is the product of hardness, cohesiveness and elasticity. Table 4 shows the relationship between some texture parameters and the popular terms used by people to describe aspects related to the texture of food products (BOURNE, 2002b).

Table 4 Relationship between texture parameters and popular terms.

Mechanical characteristics		
Primary Parameter	Secondary Parameter	Popular Terms
Hardness		Soft/Firm/Hard
Cohesiveness	Fragility	Fragile/Crunchy/Broken
	Chewability	Soft/Chewable/Resistant

Elasticity	Plastic

Source: Adapted from BOURNE (2002a).

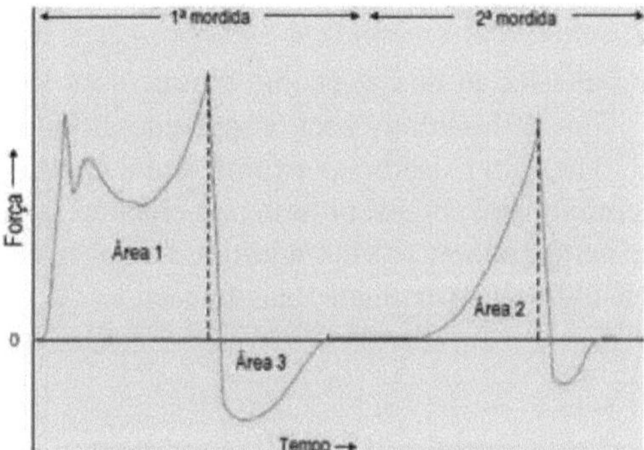

Figure 13 Representation of a typical texturometer curve.

Source: Adapted from BOURNE (2002b).

Most meat products are cooked before consumption. Physical and biochemical changes occur during the cooking process and these changes affect microbiological quality and sensory characteristics. Sensory texture is determined by the sensation in the mouth perceived during chewing, where the structure is fragmented and soaked in saliva until it is in a state suitable for swallowing. Attributes can be divided into three groups: attributes related to the breaking process (hardness, etc.); attributes related to the structure during chewing (fibrosity, elasticity, etc.) and attributes related to the end of chewing (chewing time, etc.) (BEJERHOLM and AASLYNG, 2004).

CHAPTER 3

MATERIAL AND METHODS

This chapter describes the components used in this work and the methodology used to achieve the proposed objectives. The methodology section presents how the product was made and the method used. The latter used: (a) experimental mixture planning, which was applied to the basic components, breast protein and chicken fat in order to obtain the most stable emulsion composition; (b) the addition of substitute ingredients to part of the basic components in order to evaluate their effects, as occurs in industry.

3.1 RAW MATERIALS AND INGREDIENTS

The meat patties were made with chilled chicken breast (PF), brand name Nobre, chicken fat (GF), brand name Duas Rodas, salt (S), brand name Diana, mechanically separated chicken meat (CMS), brand name Copacol, concentrated soya protein (PCS), Solae brand, sodium tripolyphosphate (STP), ICL Brasil brand and manioc starch (FM), Mundo Novo brand, as well as salts and dehydrated garlic and onion powder spices supplied by Duas Rodas Ltda.

3.2 METHODOLOGY

The work carried out was divided into two parts. One concerned the study of the basic composition on the physico-chemical characteristics of chicken mortadella and the other evaluated the effect of different ingredients on these same characteristics.

3.2.1 Study of the effect of basic chemical composition on the physico-chemical characteristics of chicken mortadella

The experimental mix design is shown in Table 5. This stage investigated the effect of protein (P), fat (G) and moisture (U) content on the emulsion stability, texture and microstructure of chicken mortadellas.

Table 5 Level and proportion of chicken mortadellas

Mixing	x1	x2	x3	U/P	PF (%)	GF (%)	A (%)
1	0,120	0,300	0,580	4,833	51,5	29,6	18,9
2	0,160	0,300	0,540	3,375	68,7	29,4	1,9
3	0,120	0,230	0,650	5,417	51,5	22,5	26,0
4	0,160	0,190	0,650	4,063	68,7	18,3	13,0
5	0,120	0,265	0,615	5,125	51,5	26,1	22,4
6	0,160	0,245	0,595	3,719	68,7	23,8	7,5
7	0,140	0,300	0,560	4,000	60,1	29,5	10,4
8	0,140	0,210	0,650	4,643	60,1	20,4	19,5
9	0,140	0,255	0,605	4,321	60,1	24,9	15,0
10	0,140	0,255	0,605	4,833	60,1	24,9	15,0

x1= protein (P); x2= fat (G); x3= moisture (U); chicken breast (PF); chicken fat (GF); water (A). Other complementary ingredients were added in the same way in all formulations.

To calculate the amounts of protein, fat and moisture in each formulation, the chemical composition of the chicken breast and fat were analysed in duplicate (AOAC, 2000). For chicken breast (PF), the results, in g-100-g^{-1}, were 23.19±0.64 protein, 1.34±0.17 fat, 74.59±1.05 moisture and 0.72g±0.14 ash; for chicken fat (GF): 0.22±0.08 protein, 99.02±0.31 fat and 0.16g±0.07 moisture. Based on these results, the chicken mortadella formulations were calculated according to the experimental mixing plan shown in Table 5, where the sum of the proportions is given by x1 + x2 + x3 = 1 (Cornell, 2002). Water (A), in the form of ice, was also used to complete the composition of the formulae and cool the dough during processing. The minimum and maximum content restrictions for each variable were defined in accordance with the limits imposed by Brazilian legislation, for which the following were established: 12.0 to 16.0% for protein (P), 19.0 to 30.0% for fat (G) and 54.0 to 65.0% for moisture (U). For every 100g of formulation, 2g of sodium chloride, 0.045g of sodium isoascorbate, 0.015g of sodium nitrite, 0.01g of garlic powder and 0.03g of onion powder were added.

3.2.2 Evaluation of the addition of different ingredients to the basic formulation of chicken mortadella

Part of the chicken breast protein and fat content was replaced in the formulation by the following ingredients: mechanically separated chicken meat (MSM), soya protein concentrate (PCS) and cassava starch (FM). The addition of the stabiliser sodium tripolyphosphate (STP) was also evaluated, according to the experimental design shown in Table 6. Below are the main characteristics of the formulations:

- addition of mechanically separated chicken meat (MSM): 10, 20, 30 and 40 per cent, corresponding to formulations 11, 12, 13 and 14;
- addition of soya protein concentrate (PCS): 1, 2, 3 and 4%, formulations 15, 16, 17 and 18;
- addition of cassava starch (FM): 1, 2, 3 and 4 per cent, formulations 19, 20, 21 and 22;
- addition of sodium tripolyphosphate (STP): 0.5 per cent with a reduction in salt (S) from 2 to 1g/100g of dough and 0.5 per cent with 2g of salt/100g of dough, corresponding to formulations 23 and 24, respectively.
- Seara brand chicken mortadella was used as a reference, containing the following ingredients in its formulation: mechanically separated poultry meat, mechanically recovered chicken meat, chicken meat, water, cassava starch, vegetable soya protein, salt, sugar, spices (garlic, white pepper and rosemary extract), stabiliser: sodium tripolyphosphate, acidulants: lactic acid, and citric acid, thickeners: carrageenan and xanthan gum, antioxidants: sodium erythorbate and ascorbic acid, preservatives sodium nitrite and sodium nitrate and natural flavourings of garlic, onion and paprika. Gluten-free. Nutritional information (100 g): Energy value: 104

kcal; Carbohydrates: 5g; Proteins: 14g; Total fats: 20g; Sodium: 1500mg. It has been estimated that this commercial chicken mortadella has approximately 60% humidity and more than 2% added salt.

Formulation 5 was selected as the base composition (protein: 12%, fat: 26.5% and moisture: 61.5%) for the study. The nutritional composition and the amount of each ingredient added were taken into account in order to maintain the chosen composition, except for the cassava starch (FM) study, where chicken fat is replaced in a 1:1 ratio, according to formulations 19, 20, 21 and 22. In this way, the approximate composition of the formulations outlined was fixed and the effects of adding these ingredients on the stability, texture and microstructure of chicken mortadellas were assessed.

In order to calculate the amounts of protein (P), fat (G) and moisture (U) for each formulation, according to the experimental design shown in Table 6, the chemical composition of the ingredients used was analysed, in $g400^{\wedge}g^{-1}$ The results are as follows: mechanically separated meat (MSM): 12.54 ± 0.47 protein, 17.10 ± 0.65 fat, 69.86 ± 1.55 moisture and $1.08g\pm0.39$ ash; for soya protein concentrate (PCS) there were 70.36 ± 1.45 protein, 3.43 ± 0.35 fat, 5.70 ± 0.74 moisture; for cassava starch there was 7.45 ± 0.68 moisture; for sodium tripolyphosphate there was 57.34 ± 1.23 phosphates, expressed as P_2O_5.

Table 6 Mortadella formulations with variations in the addition of mechanically separated chicken meat (MSM), soya protein concentrate (SPC), cassava starch (MS) and sodium tripolyphosphate (STP).

Formulation	P (%)	G (%)	U (%)	FM (%)	Total (%)	FP (%)	CMS (%)	PCS (%)	GF (%)	A (%)
5	12,0	26,5	61,5	0,0	100,0	51,50	0,00	0,00	26,10	22,40
11	12,0	26,5	61,5	0,0	100,0	46,70	10,00	0,00	23,90	19,40
12	12,0	26,5	61,5	0,0	100,0	41,30	20,00	0,00	22,30	16,40
13	12,0	26,5	61,5	0,0	100,0	35,90	30,00	0,00	20,70	13,40
14	12,0	26,5	61,5	0,0	100,0	30,40	40,00	0,00	19,10	10,50
15	12,0	26,5	61,5	0,0	100,0	48,75	0,00	1,00	25,85	24,40
16	12,0	26,5	61,5	0,0	100,0	45,70	0,00	2,00	25,90	26,40
17	12,0	26,5	61,5	0,0	100,0	42,70	0,00	3,00	25,95	28,35
18	12,0	26,5	61,5	0,0	100,0	39,70	0,00	4,00	26,00	30,30
19	12,0	25,5	61,5	1,0	100,0	51,50	0,00	0,00	25,10	22,40
20	12,0	24,5	61,5	2,0	100,0	51,50	0,00	0,00	24,10	22,40
21	12,0	23,5	61,5	3,0	100,0	51,50	0,00	0,00	23,10	22,40
22	12,0	22,5	61,5	4,0	100,0	51,50	0,00	0,00	22,10	22,40
23	12,0	26,5	61,5	0,0	100,0	51,50	0,00	0,00	26,10	22,40
24	12,0	26,5	61,5	0,0	100,0	51,50	0,00	0,00	26,10	22,40
25	12,0	24,5	61,5	2,0	100,0	21,80	40,00	3,00	17,30	15,90
MFC*	14,0	20,0	-	-	-	-	-	-	-	-

For every 100 g of formulation, 2 g of salt (S) was added, except for formulation 23, which received 1 g of salt (S). Formulations 23, 24 and 25 added 0.5 g of sodium tripolyphosphate (STP) for every 100 g of dough, in addition to the components reported in the methodology and which were kept constant for all the formulations. *MFC refers to commercial chicken mortadella, Seara brand.

3.3 MORTADELLA PREPARATION

The chicken meat emulsions, an example of which is shown in Figure 14, were processed in a cutter (Robot Coupe, R5 - Plus, France) - Figure 15(a). To carry out the mixing plan, part 1, Table 5, the process was started by comminuting the partially frozen chicken breast for 30 seconds at 1500 rpm. Then the salts and spices were added, increasing the speed to 3000 rpm, beating the dough for a further 30 seconds. Then the ice was added, maintaining the same speed for a further 60 seconds. Finally, the chicken fat was added, totalling an approximate process time of 4 minutes. For the other formulations, part 2, Table 6, a similar process was followed, with the following characteristics: the mechanically separated chicken meat (MSM) was added together with the chicken breast meat. Sodium tripolyphosphate was added along with the other salts. Soya protein concentrate (PCS) and cassava starch (FM) were added approximately 3 minutes into the process, after the emulsification of the chicken fat and protein had begun. The final temperature of the emulsification process, at the end of the comminution of the raw dough, was monitored and did not exceed 16°C. At the end of this stage of the process, around 200g of raw meat was removed to carry out the cooking loss tests.

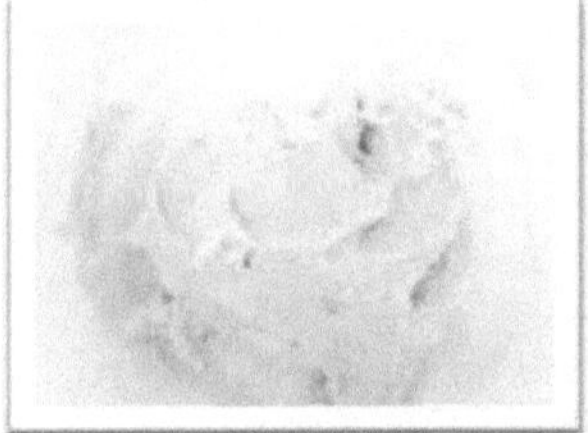

Figure 14 Raw emulsified pasta after processing.

The rest of the meat mixture, approximately 2 kg, was then stuffed, producing around 8 mortadellas with an average length of 20 cm and packed in synthetic polyethylene wrapping with an equivalent gauge of 40 mm in a Maschinenfabrik D. Nahrungsmittelmaschinen, model D-72175, Germany, as shown in Figure 15(b). Cooking was carried out in a branded oven (Eller, model I-39022, Italy), shown in Figure 15(c), programmed for three stages of time and temperature (30 minutes at 60°C, 30 minutes at 70°C and 20 minutes at 80°C, respectively). After this stage, the mortadellas were transferred to an ice bath where they remained immersed for 30 minutes, followed by storage in a cold room (Ethik Technology, model 412 TD - Brazil) at 4.0°C±0.5, (Figure 15(d)). Temperatures were monitored during the processing of the mortadella with a thermometer (Minipa, China).

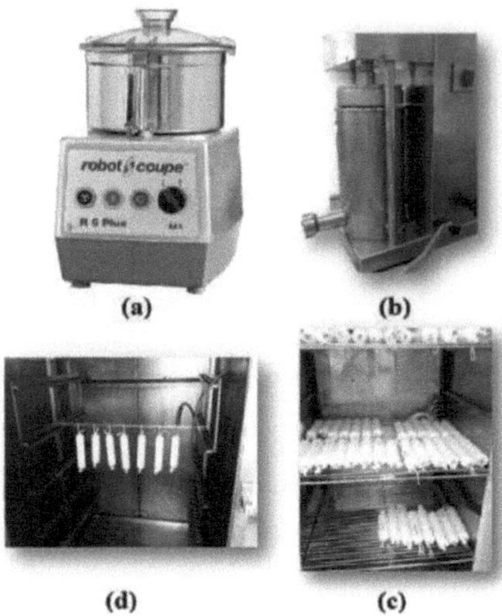

Figure 15 Equipment used in the experiments (a) Cutter; (b) Embuter; (c) Greenhouse; (d) Cold room.

3.4 PHYSICO-CHEMICAL EVALUATION

3.4.1 Cooking loss (CP)

The cooking loss (CP) analysis was based on the methodology proposed by (BARRETO, 2007). Approximately 50 grams of raw emulsified dough was weighed on an analytical balance (Mettler Toledo PB303-S, Switzerland) and packed in sealed polythene plastic bags (Selovac, Mini Jumbo model, Brazil). They were then immersed in a jacketed tank (Pemem, Brazil) at 70°C for 1 hour and cooled to room temperature, reaching an internal temperature of approximately 25°C. The final mass was weighed and the percentage of total loss over the initial mass was calculated according to Equation 1. The analysis was carried out in triplicate.

$$PC\ (\%) = \left(1 - \frac{massa\ final}{massa\ inicial}\right) \times 100 \quad (1)$$

3.4.2 Texture analysis (TPA)

Texture profile analysis (TPA) simulates the conditions under which the material is subjected to the chewing process and makes it possible to determine changes in textural properties (HERRERO et al., 2008). The main parameters determined were: firmness (N) which represents the maximum force required to compress the sample; elasticity (mm) which is the sample's ability to recover its original shape after deformation when the force is removed; cohesiveness which is the extent to which the sample could be deformed before breaking; chewability (N.mm) which represents the work required to chew the sample in order to swallow it (HORITA et al., 2011). The instrumental texture evaluations of the chicken mortadella were carried out using a texturometer (TA.XT - Plus, USA), as shown in Figure 16. The samples were standardised in the form of cylinders 40 mm in diameter and 50 mm high and analysed in triplicate. To assess the texture profile, the chicken mortadella samples were compressed to 40 per cent of their original height, with a pre-test and post-test speed of 2.0 mm/s, a test speed of 1.0 mm/s and a distance of 20 mm. The P-75 probe was used to assess hardness, elasticity, cohesiveness and chewability parameters. All the analyses were carried out with the samples at temperatures close to 20°C.

Figure 16 (a) Model box for cutting the mortadella for subsequent texture analysis; (b) texturometer (TA.XT - Plus) analysing the texture profile of a chicken mortadella.

3.4.3 Water activity and pH

The water activity, Ow, and pH determinations for the chicken mortadella were carried out using a water activity meter (AquaLab, USA) and a pH meter (Hannah, USA) with a penetration probe, respectively. The analyses were carried out in triplicate.

3.4.4 Microstructural analysis of fat

The structure of the mortadella was analysed using fluorescence microscopy. The chilled samples were fixed to a metal base using an inclusion medium for frozen samples (Tissue - T) - made up of 10.24% polyvinyl alcohol, 4.26% polyethylene glycol and 85.5% w/w of non-reactive ingredients - and placed in a cryostat (Leica Microsystems - CM 1850 UV, Germany) at a temperature of -22°C, as shown in Figure 17. After 15 minutes, when the samples were rigid, they were cut to 20^20x0.05 mm. The samples were treated with Nile Blue dye (0.01% in water) as shown in Figure 18(b) and viewed under an epifluorescence microscope (Olympus - Model Bx41, Japan), Figure 18(a) with a green filter (U-MWB2) under excitation from 460 to 490 nm and emission from 520 nm. The images were taken with a digital colour camera (Q-imaging) and captured using the Q-capture Pro 5.1 software (Q-imaging).

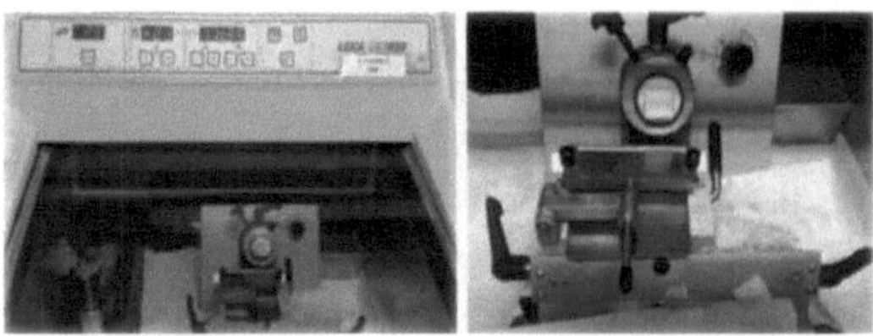

Figura 17   Cryostat used to prepare chicken mortadella samples for subsequent microscopy analysis.

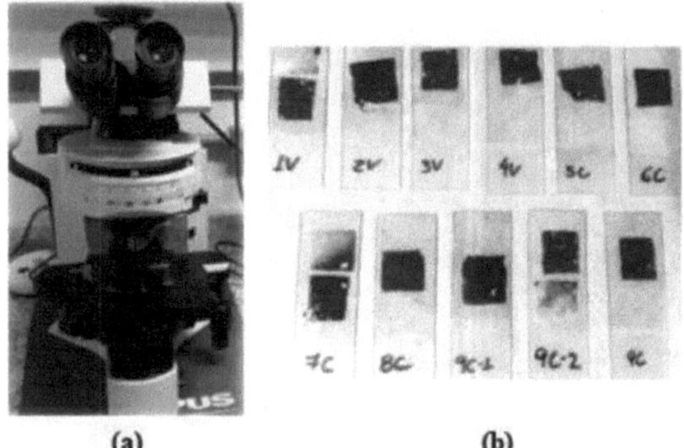

(a)   **(b)**

Figura 18   (a) Olympus microscope (model Bx41) used to analyse fluorescence microscopy. (b) slides with different formulations of chicken mortadella stained with nile blue dye.

3.5 STATISTICAL EVALUATION

The results obtained were statistically analysed using analysis of variance (ANOVA) and Tukey's test at a 5% significance level using Statistica software®. A three-component restricted mixture experimental design was used.

CHAPTER 4

RESULTS AND DISCUSSION

4.1 STUDY OF THE EFFECT OF BASIC CHEMICAL COMPOSITION ON THE PHYSICOCHEMICAL CHARACTERISTICS OF CHICKEN MORTADELLA

The formulations used to prepare the chicken mortadellas, with compositions ranging from 12 to 16 per cent protein, 19 to 30 per cent fat and 54 to 65 per cent moisture, as shown in Table 5, show that the levels of protein, fat and moisture had a significant effect on the cooking losses and texture profile of the chicken mortadellas (Table 7).

The temperatures monitored during processing show similar behaviour, as shown in Figure 19. On average, the dough temperature at the end of the emulsification process was 10.23±5.51°C. These processing conditions are in line with those indicated by PEARSON and GILLETT (1996) whose value is below 15°C as ideal for emulsification processes of finely comminuted products, as these conditions prevent the fat from coalescing.

At the end of the first cooking stage, the temperature in the centre of the mortadella was 47.80±2.25°C. According to the literature (Pearson and Gillett, 1996) the process of denaturation and gelation of myosin begins at approximately 40°C, which contributes to the formation of the protein matrix. At the end of the second and third stages, temperatures in the centre of the mortadellas of 64.90±1.52°C and 74.0±1.82°C, respectively, were measured. According to the same authors, temperatures between 65 and 70°C promote the coagulation of proteins with the appearance of texture. It can be seen that 72°C is the internal temperature indicated for pasteurising the product and the end of cooking.

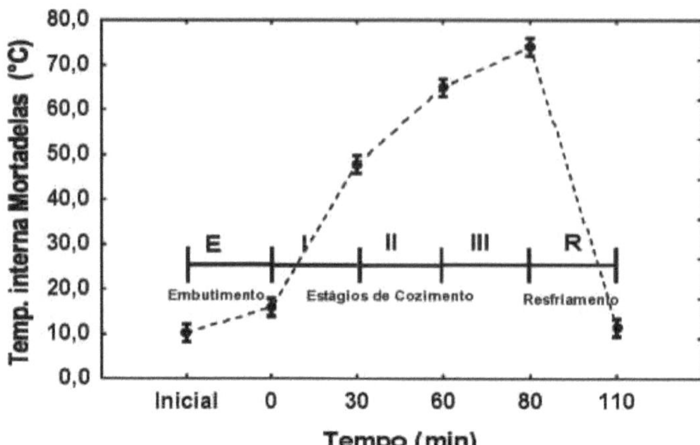

Figura 19   Temperature measured in the centre of the product at different processing stages of chicken mortadella.

Source: the author.

During the cutter comminution process of the raw chicken pasta, homogeneous emulsions with a light colour were formed, close to the colour of the cooked chicken meat. In general, the compositions with lower moisture/protein ratios showed chicken pasta with visually higher observed viscosities due to a higher protein concentration.

In this thermodynamically unstable environment, the emulsion stabilisation process begins with the formation of an emulsifying film around the fat particles. In this case, there is a realignment of the proteins that position their hydrophobic region with the oil phase and their hydrophilic region with the aqueous phase (LAM and NICKERSON, 2013).

In the cooking loss tests, the total mass of liquid resulting from the destabilisation of the emulsion was quantified for each formulation. The characteristics of these liquids were visually observed: transparency and absence of suspended particles and apparent fat.
The texture, cohesiveness and elasticity parameters are related to the forces involved in the internal bonds of the mortadellas. However, the results for these experiments were very close, and considering the experimental error, it was not possible to observe a clear trend regarding the influence of the variables studied. Also, the results for the influence of a_w and pH did not differ significantly (Table 7).

Table 7 Average values for cooking losses, water activity, pH, firmness, elasticity, cohesiveness and chewiness of chicken mortadellas.

F*	Cooking losses (%)	a_w	pH	Firmness (N)	Elasticity (mm)	Cohesiveness	Chewability (N-mm)
1	9,19 ± 0,70[a]	0,977 ± 0,005[a]	6,02 ± 0,05[a]	71,843 ± 2,391[f]	0,859 ± 0,016b	0.674 ± 0.020[abcdef]	41,592 ± 2,718[f]
2	2,50 ± 0,29[d]	0,981 ± 0,006[a]	6,04 ± 0,02[a]	153,322 ± 3,755[a]	0,855 ± 0,004[b]	0.659 ± 0.004[bcdef]	86,367 ± 2,594[a]
3	9,78 ± 0,10[a]	0,972 ± 0,002[a]	5,95 ± 0,01[a]	54,698 ± 2,309[h]	0.885 ± 0.003[ab]	0.711 ± 0.007[abcd]	34,406 ± 1,331[g]
4	5,90 ± 0,67[c]	0,980 ± 0,004[a]	6,00 ± 0,08[a]	107,940 ± 2,963[c]	0,893 ± 0,007[a]	0.694 ± 0.010[abcdef]	66,907 ± 2,256[c]
5	9,60 ± 0,34[a]	0,977 ± 0,005[a]	5,92 ± 0,03[a]	63,095 ± 1,736[g]	0.876 ± 0.028[ab]	0.714 ± 0.032[abcd]	39,450 ± 1,967[f]
6	4,14 ± 0,47[c]	0,973 ± 0,001[a]	6,04 ± 0,01[a]	121,025 ± 0,607[b]	0.882 ± 0.005[ab]	0.715 ± 0.007[abcd]	76,331 ± 1,252[b]

F							
7	$4{,}28 \pm 0{,}55^c$	$0{,}979 \pm 0{,}002^a$	$5{,}98 \pm 0{,}03^a$	$120{,}995 \pm 3{,}921^b$	$0{,}856 \pm 0{,}009^b$	0.652 ± 0.018^{bdef}	$67{,}503 \pm 0{,}809^c$
8	$9{,}01 \pm 0{,}51^a$	$0{,}978 \pm 0{,}003^a$	$5{,}94 \pm 0{,}06^a$	$80{,}286 \pm 1{,}707^e$	0.879 ± 0.008^{ab}	0.708 ± 0.009^{abcd}	$48{,}981 \pm 2{,}469^e$
9	$6{,}28 \pm 0{,}36^b$	$0{,}974 \pm 0{,}005^a$	$6{,}03 \pm 0{,}08^a$	$93{,}025 \pm 2{,}504^d$	0.878 ± 0.002^{ab}	0.677 ± 0.013^{abcdef}	$55{,}221 \pm 0{,}532^d$
10	$6{,}68 \pm 0{,}43^b$	$0{,}974 \pm 0{,}002^a$	$6{,}02 \pm 0{,}08^a$	$90{,}918 \pm 2{,}014^d$	0.881 ± 0.005^{ab}	0.702 ± 0.017^{abcde}	$56{,}208 \pm 2{,}660^d$

Averages refer to three repetitions of analyses for the same sample. Different letters within columns indicate significant differences ($p < 0.05$). *F= formulation number.

Source: the author.

In the region designated as low stability (Figure 20), chicken mortadella formulations with higher cooking losses (between 5.90 ± 0.67 and 9.78 ± 0.10) were observed, according to Table 7. There was a trend towards higher cooking losses as the moisture/protein ratio increased (Figure 21). The same compositions had lower firmness and chewiness values in terms of texture, ranging from 54.698 ± 2.309 to 107.947 ± 2.616 N and from 34.406 ± 1.331 to 66.183 ± 2.221 N-mm (Figures 22 and 24, respectively). It was probably the lower concentration of protein present in the protein matrix that influenced the modification of the textural properties and stability of the product. The firmness and chewiness values decreased as the moisture/protein ratio increased (Figures 23 and 25). The formation of larger fat globules, such as those seen in Figure 26, indicates a process of fat coalescence due to the rupture or formation of the ineffective emulsifying film around them.In this region designated as the high stability region (Figure 20), chicken mortadella formulations with reduced cooking losses were observed (between $2.50\pm0.29\%$ and $4.28\pm0.55\%$). As for texture (Figures 22 and 24), these compositions showed high firmness and chewiness results, with values ranging from 120.995 ± 3.921 to 153.323 ± 3.755 N, and from 67.503 ± 0.809 to 86.367 ± 2.594 N-mm, respectively. Compositions with a high protein content and low moisture content were observed in this area. In this sense, the homogeneity and distribution of the smaller fat globules can be attributed to an adequate protein content to emulsify this quantity of lipids, which resulted in greater emulsion stability (Figure 27). On the other hand, it is known that, theoretically, the presence of fat allowed the lipophilic portion of the proteins to unfold, associating with the fat, while the hydrophilic portions remained in the aqueous phase (PEARSON and GILLETT, 1996).

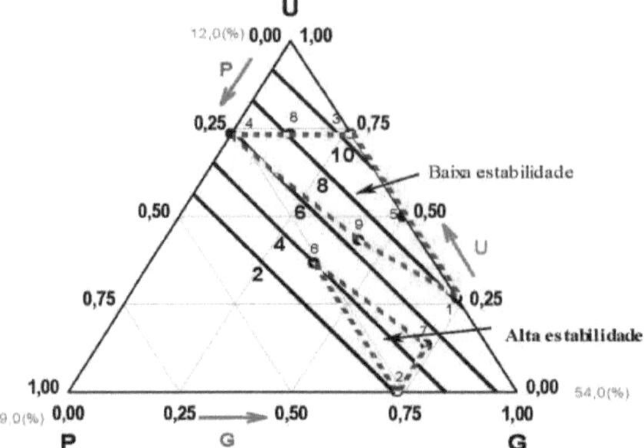

Figura 20   Graph of the effects of protein, fat and moisture content on the cooking losses (%) of chicken mortadella, $R^2 = 0.906$.

Source: the author.

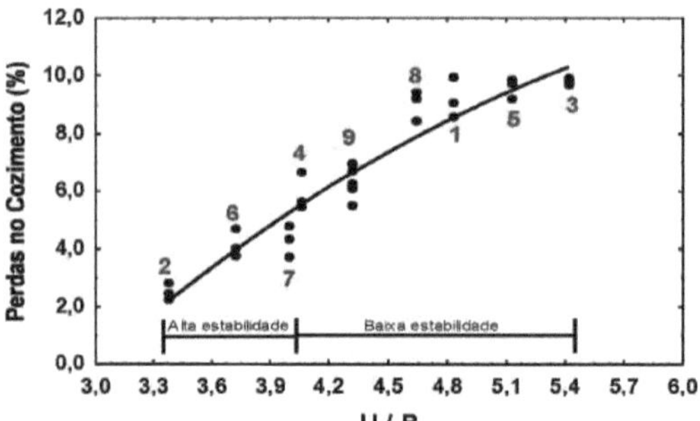

Figura 21   Graph of cooking losses (%) as a function of the (Moisture/Protein) ratio, $R^2 = 0.9471$.

Source: the author.

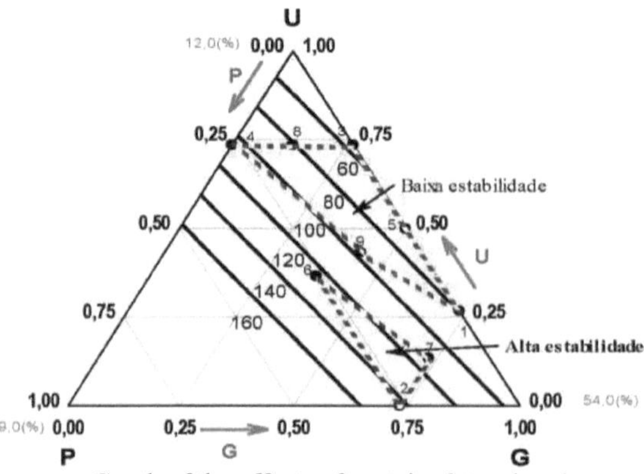

Figura 22   Graph of the effects of protein, fat and moisture content on the Firmness (N) of chicken mortadellas. $R^2 = 0,9682$.

32

Source: the author.

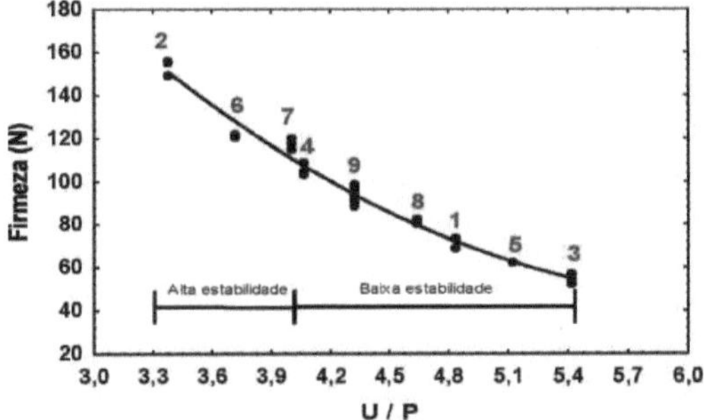

Figura 23   Graph of Firmness (N) as a function of the ratio (Moisture/Protein), $R^2 = 0.9766$.

Source: the author.

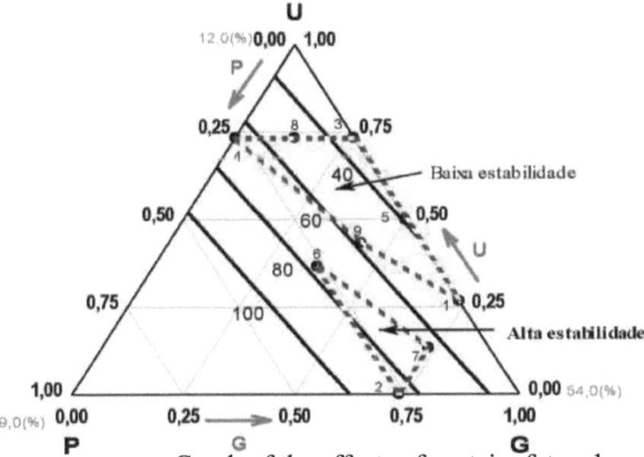

Figura 24   Graph of the effects of protein, fat and moisture content on the Chewiness (N-mm) of chicken mortadellas, R2 = 0.9762.

Source: the author.

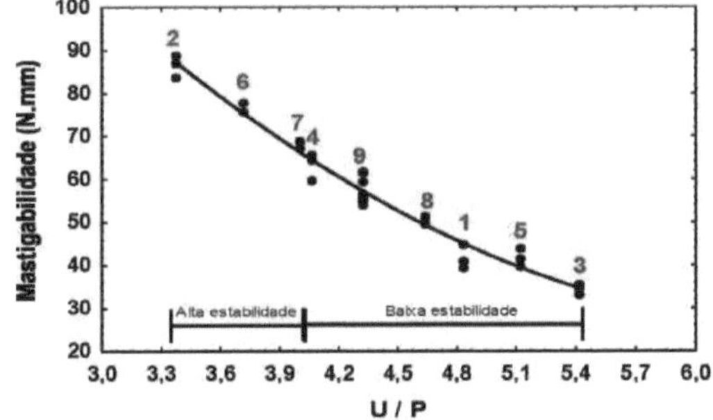

Figura 25   Graph of Chewiness (N^mm) as a function of the (Moisture/Protein) ratio, $R^2 = 0.9762$.

Source: the author.

33

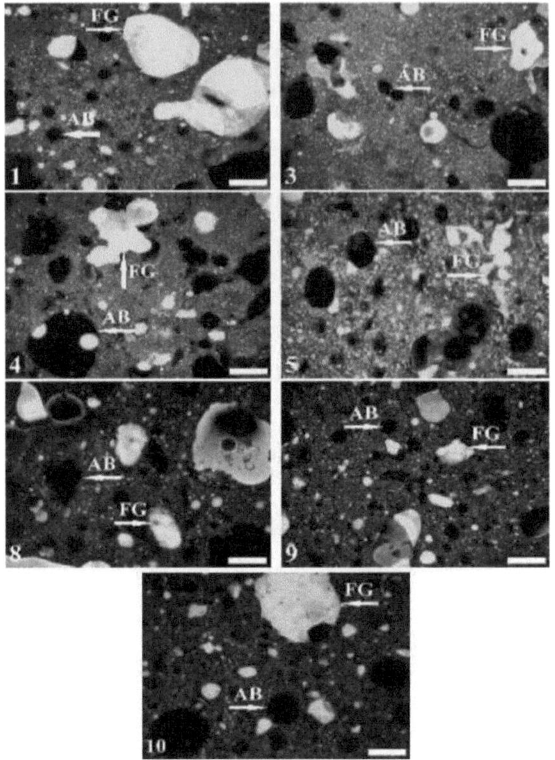

Figura 26 Fluorescence microscopy of chicken mortadellas. Region of lower emulsion stability. (1) 12% P; 30% G; 58% U, (3) 12% P; 23% G; 65% U, (4) 16% P; 19% G; 65% U, (5) 12% P; 26.5% G; 61.5% U, (8) 14% P; 21% G; 65% U, (9) and (10) 14% P; 25.5% G; 60.5% U. FG-Fat globules; AB-Air bubbles. Bar: 200 gm.

Source: the author.

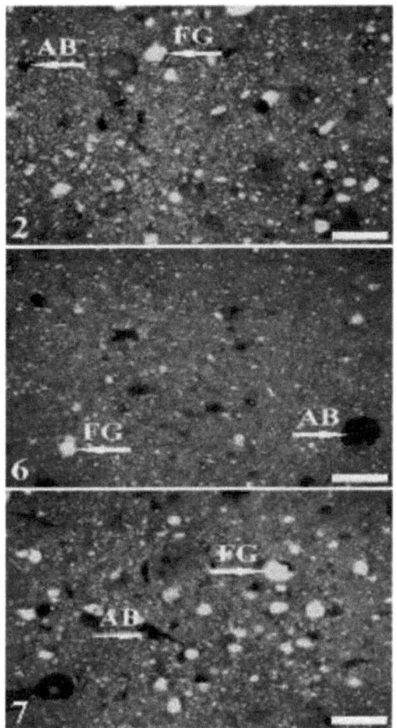

Figura 27   Fluorescence microscopy of chicken mortadellas. Region of greatest emulsion stability. (2) 16% P; 30% G; 54% U, (6) 16% P; 24.5% F; 59.5% M, (7) 14% P; 30% F; 56% M FG-Fat globules; AB-Air bubbles. Bar: 200 gm.

Source: the author.

4.2 EVALUATION OF THE ADDITION OF INGREDIENTS TO THE BASIC FORMULATION OF CHICKEN MORTADELLA

The results of adding ingredients will be compared to a formulation with low stability. Formulation 5 was chosen as the reference for comparison because it had one of the highest cooking loss values, and was no different from the others in the same loss range (Formulations 1, 3 and 8). Furthermore, among these formulations, another factor that influenced the choice of formulation number 5 was the fact that it had a composition with a minimum of protein (emulsifier) and an average amount of fat (dispersed phase). In this way, the replacement of the basic raw materials with additional ingredients could be more strongly emphasised.

4.2.1 Addition of mechanically separated chicken meat (MSM)

Formulations 11, 12, 13 and 14, with 10, 20, 30 and 40 per cent mechanically separated chicken meat (MSC) added, showed cooking losses of between 6.77±0.14 per cent and 7.39±0.17 per cent (Table 8). According to Figure 28(a), increasing the amount of CMS did not significantly alter the cooking losses of the mortadellas (p<0.05), except when compared to the basic formulation (number 5) which showed the highest loss (9.60 ± 0.34%) without CMS. In this case, the good emulsification capacity of the SMC may have favoured lower cooking losses (JIMÉNEZ COLMENERO, 2004).

With regard to pH, there was a significant increase (p<0.05) with the increase in the CMS in the formulation, Figure 28(b), showing values between 5.98 ± 0.02 and 6.10 ± 0.01, which can be explained by the (CMS) having a higher pH, due to its specific physico-chemical and microbiological characteristics and composition. It is also possible that the pH influenced the reduction in cooking losses compared to formulation 5, used as a reference.

In terms of texture, the same formulations had firmness and chewiness values varying between 59.718±4.941 and 63.343±4.772 N, and 35.634±1.661 and 42.636±4.315 N-mm, respectively, which did not differ significantly (p<0.05), even when compared to formulation 5 (Figure 28(c)). The addition of up to 40 per cent CMS kept the formulations with a similar microscopic distribution of fat globules (Figure 29), differing only slightly from formulation 5, which does not have mechanically separated chicken meat (CMS) in its formulation.

Thus, taking into account factors such as availability and lower cost, the use of 40% mechanically separated chicken meat (MSC) in the formulation of chicken mortadella reduced cooking losses and maintained the product's firmness and chewiness values. As for the increase in pH value, the use of MSM may favour a reduction in the product's shelf life due to the physicochemical and microbiological characteristics inherent in the production process.

Table 8 Average values for cooking losses, water activity, pH, firmness, elasticity, cohesiveness and chewiness of chicken mortadellas with added mechanically separated meat (MSM).

F*	Cooking losses (%)	a$_w$	pH	Firmness (N		Elasticity (mm)	Cohesive ness	Chewability (N.mm)
5	9,60 ± 0,34[a]	0,977 ± 0,005[a]	5,92 ± 0,03[c]	63,095 1,736[a]	±	0,876 ± 0,028[a]	0,714 ± 0,032[a]	39,450 ± 1,967[a]
11	7,39 ± 0,17[b]	0,981 ± 0,006[a]	5,98 ± 0,02[b]	63,343 4,772[a]	±	0,895 ± 0,002[a]	0,709 ± 0,005[a]	40,208 ± 3,315[a]
12	7,33 ± 0,92[b]	0,972 ± 0,002[a]	6,08 ± 0,01[a]	59,718 4,941[a]	±	0,867 ± 0,016[a]	0,691 ± 0,041[a]	35,634 ± 1,661[a]
13	6,77 ± 0,14[b]	0,980 ± 0,004[a]	6,08 ± 0,02[a]	62,867 2,361[a]	±	0,878 ± 0,015[a]	0,720 ± 0,016[a]	39,733 ± 1,182[a]
14	7,06 ± 0,19[b]	0,977 ± 0,005[a]	6,10 ± 0,01[a]	63,276 2,901[a]	±	0,901 ± 0,013[a]	0,746 ± 0,035[a]	42,636 ± 4,315[a]

Averages refer to three repetitions for the same sample. Different letters in the columns indicate significant differences (p <0.05). *F= formulation number. Source: the author.

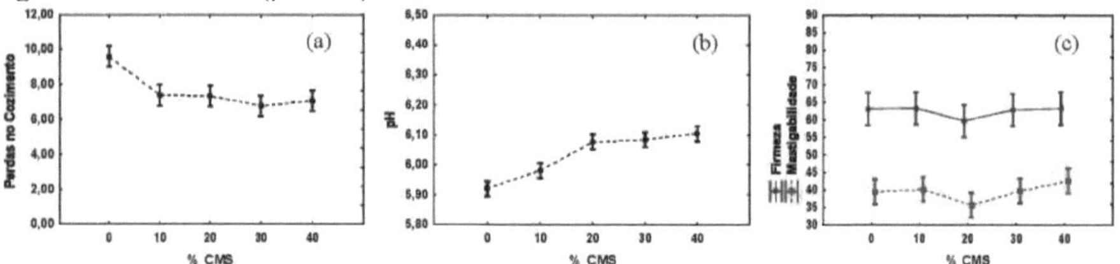

Figure 28 Graphs of (a) cooking losses, (b) pH and (c) firmness and chewiness as a function of the variation in mechanically separated chicken meat (MSM).
Source: the author.

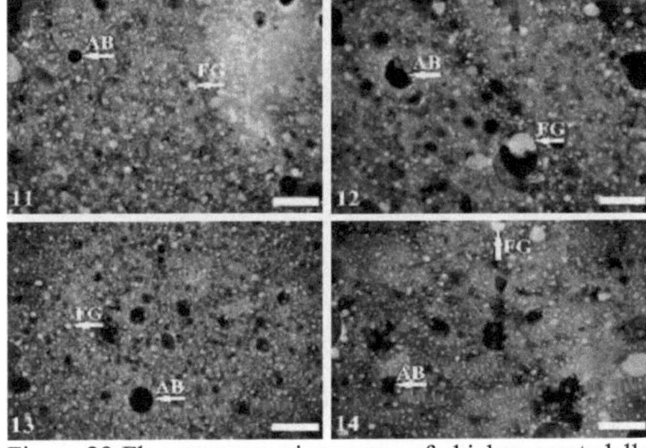

Figure 29 Fluorescence microscopy of chicken mortadellas. Formulations containing 0, 10, 20, 30 and 40 per cent mechanically separated meat (MSM), respectively. FG-Fat globules; AB-Air bubbles. Bar: 200jim.

Source: the author.

4.2.2 Addition of soya protein concentrate (PCS)

The formulations with 1, 2, 3 and 4 % (trials 15, 16, 17 and 18) of soya protein concentrate (PCS) showed cooking losses of between 3.84 ± 0.33 and 8.07 ± 0.14, differing significantly from each other ($p < 0.05$), even when compared to formulation 5, with 9.60 ± 0.34% losses (Table 9). There was a downward trend in cooking losses with the addition of soya protein concentrate (PCS), Figure 30(a). The decrease may be due to the interaction between soya protein and water. This can be explained by the variability of the molecular weights of this protein, as well as its solubility characteristics in water as well as in saline solution.

As for pH, there was no significant difference ($p < 0.05$) with the increase in soya protein concentrate (SCP) content, Figure 30(b), showing values between 6.05±0.01 and 6.09±0.01, except when compared to formulation 5 (5.92±0.03), used as a reference, without SCP in its composition.

In terms of texture, the same formulations with added PCS showed firmness and chewiness values varying between 52.824±2.977 and 67.605±1.092 N, and 34.411±2.079 and 43.900±0.585 N.mm, respectively, according to Table 9. The addition of a lower soya protein content, 1 and 2 per cent of PCS (formulations 15 and

16), respectively, showed no significant differences between them. However, both firmness and chewiness decreased when compared to those with no and 3 and 4 per cent PCS, respectively (Figure 30(c)). Formulation 18, with 4% PCS, was significantly higher than all the others. The addition of up to 4% PCS kept the emulsions structurally similar, as can be seen in the microscopy (Figure 31), where the fat globules showed homogeneous distribution and size, differing only slightly from formulation 5, with no soya protein concentrate (PCS).

The addition of between 3 and 4% soya protein concentrate (PCS) reduced cooking losses and significantly increased the firmness and chewiness of the chicken mortadella evaluated. In this case, the higher concentration of this type of protein in the formulation, as a substitute for meat protein, may have favoured improving the stability of the emulsion and the textural characteristics of the product.

Table 9 Average values for cooking losses, water activity, pH, firmness, elasticity, cohesiveness and chewiness of chicken mortadellas with added soya protein concentrate (PCS).

F*	Cooking losses (%)	aw	PH	Firmness (N	Elasticity (mm)	Cohesive ness	Chewability (N.mm)
5	9,60 ± 0,34[a]	0,977 ± 0,005[a]	5,92 ± 0,03b	63.095 ± 1.736[bcd]	0,876 ± 0,028[a]	0,714 ± 0,032[a]	39.450 ± 1.967b[c]
15	8,07 ± 0,14[b]	0,978 ± 0,002[a]	6,05 ± 0,01[a]	52,924 ± 1,218[b]	0,895 ± 0,006[a]	0,741 ± 0,011[a]	35.116 ± 0.302[bd]
16	6,27 ± 0,26[c]	0,974 ± 0,003[a]	6,07 ± 0,02[a]	52,824 ± 2,977[b]	0,899 ± 0,004[a]	0,725 ± 0,004[a]	34.411 ± 2.079[bd]
17	4,21 ± 0,37[d]	0,977 ± 0,002[a]	6,09 ± 0,01[a]	59.775 ± 2.004[bcd]	0,896 ± 0,002[a]	0,728 ± 0,019[a]	39.009 ± 1.966[bc]
18	3,84 ± 0,33[d]	0,979 ± 0,001[a]	6,05 ± 0,02[a]	67,605 ± 1,092[a]	0,876 ± 0,002[a]	0,741 ± 0,006[a]	43,900 ± 0,585[a]

Averages refer to three repetitions of analyses for the same sample. Different letters within columns indicate significant differences (p <0.05). *F= formulation number. Source: the author.

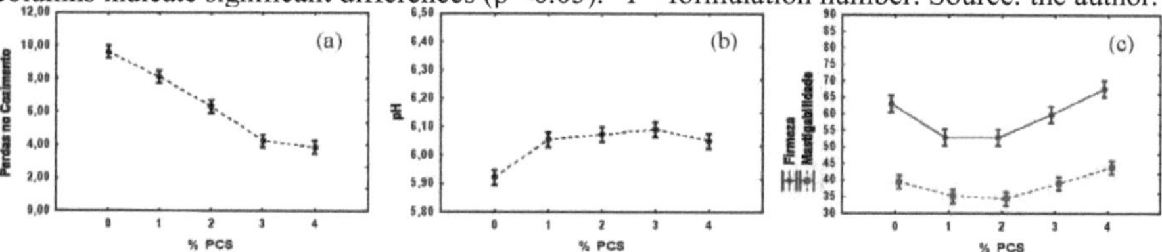

Figure 30 Graphs of (a) cooking losses, (b) pH and (c) firmness and chewiness as a function of the variation in soya protein concentrate (PCS).

Source: the author.

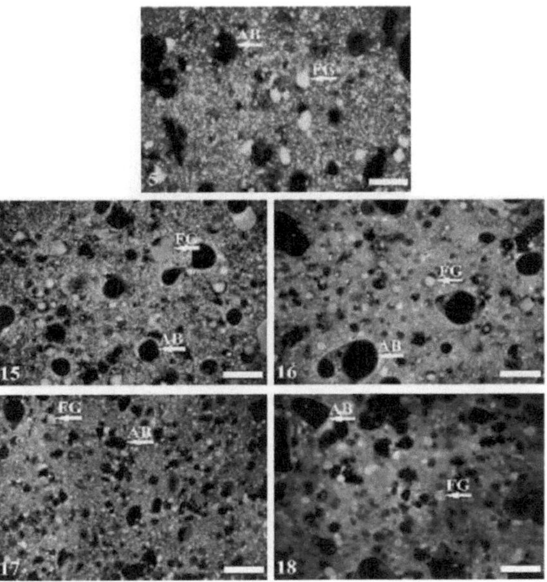

Figure 31 Fluorescence microscopy of chicken mortadellas. Formulations 5, 15, 16, 17 and 18, containing 0, 1, 2, 3 and 4% soya protein concentrate (PCS), respectively. FG-Fat globules; AB-Air bubbles. Bar: 200|j,m.

Source: the author.

4.2.3 Addition of cassava starch (FM)

The formulations with 1, 2, 3 and 4 per cent cassava starch (trials 19, 20, 21 and 22) showed cooking losses of between 2.14 ± 0.38 and 7.12 ± 0.58, differing significantly from each other (p<0.05), even when compared to formulation 5, with 9.60 ± 0.34 per cent losses (Table 10). There was a downward trend in cooking losses with the addition of cassava starch (FM), Figure 32(a). However, the formulations with 3 and 4 per cent (21 and 22) did not differ significantly from each other, both showing reduced cooking losses. The addition of starch to cooked products causes the water to be strongly bound during cooking, explaining the reduction in cooking losses.

The pH increased slightly with the addition of cassava starch, Figure 32(b), showing values between 5.92 ± 0.03 and 6.15 ± 0.04.

As for texture, the FM-added formulations showed firmness and chewiness values varying between 62.804±1.899 and 77.770±7.339 N, and 40.601±0.912 and 49.242±3.839 N-mm, respectively, as shown in Table 11. There was a tendency for these two parameters to increase with the addition of cassava starch (FM). The addition of up to 4% FM kept the emulsions structurally similar, as can be seen in the microscopy (Figure 33), where the fat globules showed homogeneous distribution and sizes, differing only from formulation 5, with no cassava starch (FM).

As a result, it was felt that the addition of 3% cassava starch was adequate to reduce cooking losses and slightly increase firmness and chewiness, improving the stability of the product.

Table 10 Average values for cooking losses, water activity, pH, firmness, elasticity, cohesiveness and chewiness of chicken mortadellas with added cassava starch (FM).

F*	Cooking losses (%)	a_w	PH	Firmness (N	Elasticity (mm)	Cohesiveness	Chewability (N.mm)
5	9,60 ± 0,34[a]	0,977 ± 0,005[a]	5,92 ± 0,03[b]	63,095 ± 1,736[c]	0,876 ± 0,028[a]	0,714 ± 0,032[a]	39,450 ± 1,967[b]
19	7,12 ± 0,58[b]	0,977 ± 0,001[a]	5,99 ± 0,03[b]	62,804 ± 1,899[c]	0,890 ± 0,003[a]	0,709 ± 0,020[a]	40,601 ± 0,912[b]
20	5,01 ± 0,15[c]	0,974 ± 0,001[a]	6,13 ± 0,03[a]	73.171 ± 1.568[ab]	0,896 ± 0,013[a]	0,719 ± 0,013[a]	47.137 ± 0.673[ac]
21	2,14 ± 0,38[d]	0,978 ± 0,001[a]	6,15 ± 0,04[a]	66.297 ± 1.831[bc]	0,904 ± 0,006[a]	0,726 ± 0,020[a]	43.516 ± 0.703[abc]
22	2,69 ± 0,60[d]	0,975 ± 0,003[a]	6,10 ± 0,04[a]	77.770 ± 7.339[ab]	0,895 ± 0,004[a]	0,708 ± 0,020[a]	49.242 ± 3.839[ac]

Averages refer to three repetitions of analyses for the same sample. Different letters within columns indicate significant differences ($p < 0.05$). *F= formulation number. Source: the author.

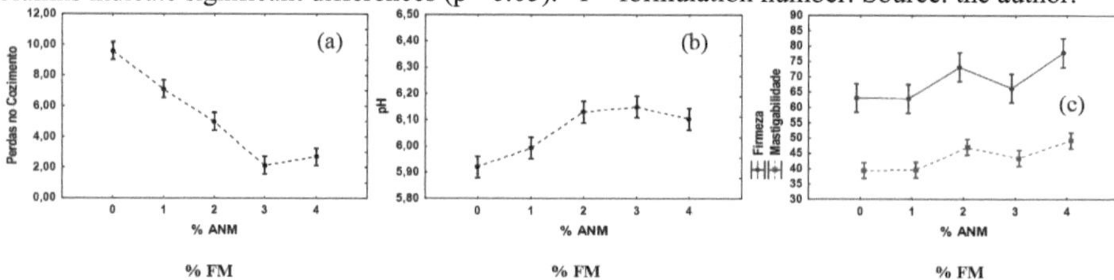

Figure 32 Graphs of (a) cooking losses, (b) pH and (c) firmness and chewiness as a function of cassava starch (FM) variation.
Source: the author.

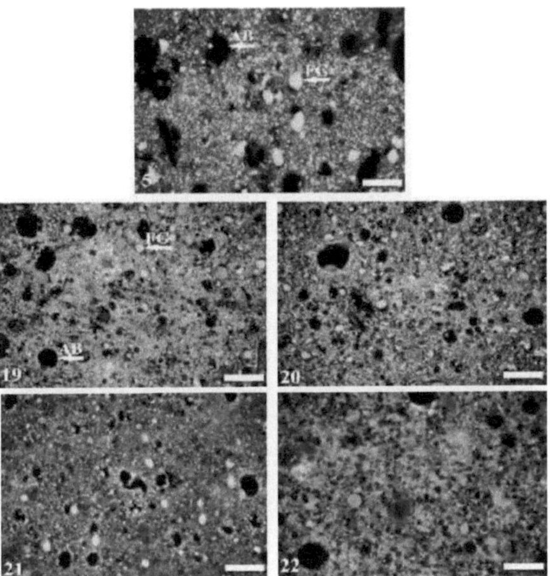

Figure 33 Fluorescence microscopy of chicken mortadellas. Formulations 5, 19, 20, 21 and 22, containing 0, 1, 2, 3 and 4% cassava starch (FM), respectively. FG-Fat globules; AB-Air bubbles. Bar: 200jim.

Source: the author.

4.2.4 Addition of sodium tripolyphosphate (STP)

Formulations 23 and 24 with added sodium tripolyphosphate (STP) showed cooking losses of between 2.64±0.56 and 3.85±0.15, differing significantly from each other, even when compared to formulation 5 (9.60 ± 0.34%) in Table 11. The sharp drop in this parameter (Figure 35(a)) may be partly due to the increase in pH (Figure 35b), which shifts from the pH of the isoelectric point of the proteins, allowing more water to bind to these molecules. However, the water retention capacity cannot be explained solely by the increase in pH, because when the increase occurs due to other factors, there is no proportional increase. The phosphates, together with the salt (S), increase the interfilament spaces, expanding the myofibrils and improving the water retention of the meat. The action of this additive is so significant that even with the reduction in salt (S), formulation 23, there was a reduction in cooking losses. In general, phosphates can dissociate the actomyosin complex. Studies have shown that the addition of phosphates together with salt, by modifying the myofibrillar structure, breaks the bonds between the actin and myosin filaments, increasing water retention capacity. To a lesser extent, the complexation of the Ca^{++} and Mg^{++} ions present in the meat form bridges between the negative charges of the proteins, which, when broken by the action of the phosphates, can improve the water retention capacity (LEMOS et al., 2011).

In terms of texture, the firmness values of the formulations ranged from 49.787±1.404 to 58.118 ±2.605±1.736 N, differing significantly (p<0.05) when compared to formulation 5 (63.095±1.736 N) in Table 11. However, the addition of sodium tripolyphosphate reduced the firmness of the chicken mortadellas (Figure 35c). The greater retention probably made the products softer due to the plasticising capacity of the water. As for chewiness, which showed values between 34.487±0.583 and 39.650±1.035 N.mm, these were also significantly different from each other, with the exception of formulations 5 (39.450 ± 1.967 N.mm) and 24 (39.650±1.035 N-mm). With the reduction in salt, formulation 23, there was a reduction in firmness and chewiness, as the extraction of soluble proteins in saline solution was probably impaired.

The addition of 0.5 g of sodium tripolyphosphate (STP) per 100 g of dough kept the emulsions structurally similar, as can be seen in the microscopy (Figure 34), where the fat globules showed homogeneous distribution and size. However, there were significant differences between formulations 5, 23 and 24.

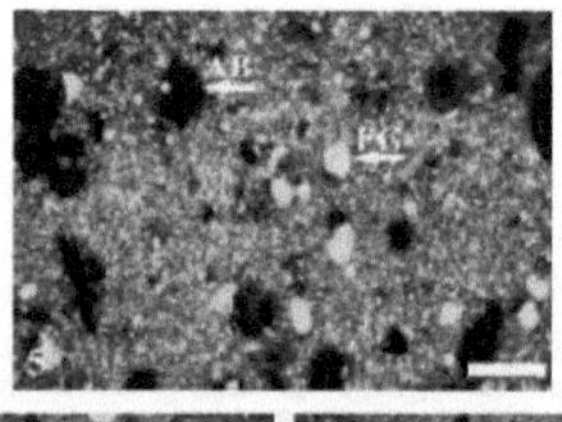
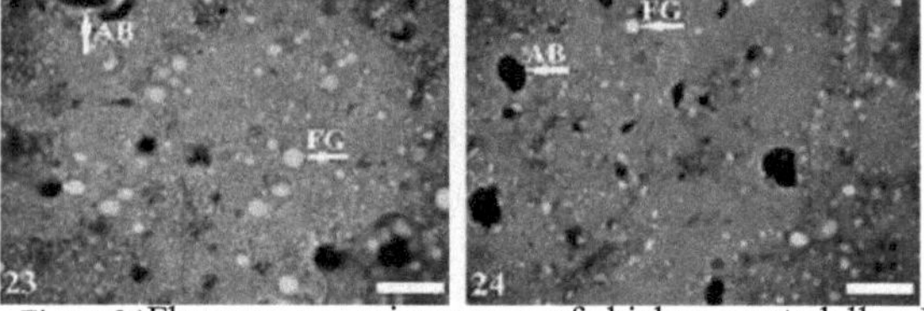

Figura 34Fluorescence microscopy of chicken mortadellas. Formulations 23 and 24 containing 1 and 2 g of salt (S) per 100 g of dough, respectively, with the addition of 0.5 g of sodium tripolyphosphate per 100 g of dough. Formulation 5 without the addition of (STP). FG-Fat globules; AB-Air bubbles. Bar: 200 gm.

Source: the author.

Table 11 Average values for cooking losses, water activity, pH, firmness, elasticity, cohesiveness and chewiness of chicken mortadellas with added sodium tripolyphosphate (STP).

F*	Cooking losses (%)	a_w	PH	Firmness (N)	Elasticity (mm)	Cohesive ness	Chewability (N.mm)
5	9,60 ± 0,34[a]	0,977 ± 0,005[a]	5,92 ± 0,03[c]	63,095 ± 1,736[a]	0,876 ± 0,028[a]	0,714 ± 0,032[a]	39,450 ± 1,967[a]
23	3,85 ± 0,15[b]	0,985 ± 0,002[a]	6,16 ± 0,02[b]	49,787 ± 1,404[c]	0,901 ± 0,003[a]	0,770 ± 0,031[a]	34,487 ± 0,583[b]
24	2,64 ± 0,56[c]	0,980 ± 0,002[a]	6,27 ± 0,02[a]	58,118 ± 2,605[b]	0,907 ± 0,002[a]	0,753 ± 0,023[a]	39,650 ± 1,035[a]

Averages refer to three repetitions of analyses for the same sample. Different letters within columns indicate significant differences (p <0.05). *F= formulation number. Source: the author.

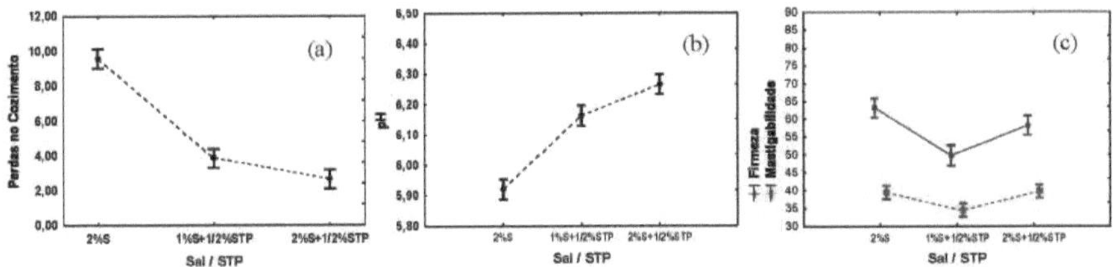

Figura 35   Graphs of (a) cooking losses, (b) pH and (c) firmness and chewiness as a function of sodium tripolyphosphate (STP) variation.

Source: the author.

4.3 THEORETICAL-EXPERIMENTAL FORMULATION PROPOSAL VERSUS COMMERCIAL CHICKEN BOLOGNA (CBM)

Based on the results obtained, a theoretical-experimental proposal (Formulation 25) was drawn up for comparison with Seara brand commercial chicken mortadella (MFC) (Table 6). The construction of this composition took into account the results already obtained for all the ingredients tested. These were:

- 40% mechanically separated chicken meat was added due to its widespread use in the mortadella processing industry and lower cost compared to chicken breast meat. In addition, the experiments carried out showed a slight reduction in cooking losses and maintenance of the product's textural parameters.
- The main objective of using 3% soya protein concentrate was to reduce cooking losses in the products, as demonstrated by the fact that this ingredient is highly used in the industry to complement and replace meat protein.
- the decision was made to use 2 per cent cassava starch (FM). It was noted that this ingredient reduced cooking losses and, as it would be used together with the soya protein concentrate, it was decided not to add more, even though it is widely used in the industry within the limits of its legal use. The high fat content of this formulation was also taken into account.
- The addition of 0.5 g of sodium tripolyphosphate (STP) per 100 g of dough was maintained due to its high performance in reducing cooking losses and improving emulsion stability.
- the amount of salt added was kept at 2g per 100g of dough, due to the commercial formulation (MFC) possibly being higher due to the amount of sodium indicated in its nutritional table.

When comparing mortadella 25 and MFC, significant differences were observed with regard to pH, firmness and elasticity. The parameters water activity, cohesiveness and chewiness did not differ (Table 12). The differences in pH may be due to the different ingredients and raw materials used, including the acidulants lactic acid and citric acid present in the commercial formulation. As for firmness, the higher protein content of the MFC may have positively influenced this parameter (Figure 36), explaining its higher value compared to formulation 25. However, the chewiness parameter was similar between the two products. It was observed that the commercial chicken mortadella (MFC) had larger fat globules (Figure 37), possibly due to the lack of comminution and not the instability of the emulsion. There was also less air in the (MFC), probably due to the better filling process, usually carried out in vacuum equipment in the industry.

Table 12 Average values for cooking losses, water activity, pH, firmness, elasticity, cohesiveness and chewiness, for comparison of chicken mortadella 25 and MFC.

F*	aw	pH	Firmness (N)	Elasticity (mm)	Cohesiveness	Chewability (N.mm)
25	0,977 ± 0,002[a]	6,43 ± 0,03[a]	65,841 ± 2,843[b]	0,906 ± 0,007[a]	0,753 ± 0,036[a]	44,981 ± 3,961[a]
MFC	0,976 ± 0,001[a]	6,12 ± 0,03[b]	76,851 ± 2,796[a]	0,874 ± 0,006[b]	0,732 ± 0,021[a]	49,195 ± 2,877[a]

*Formulation. Averages refer to three repetitions for the same sample. Different letters in the columns indicate significant differences (p <0.05). Source: the author.

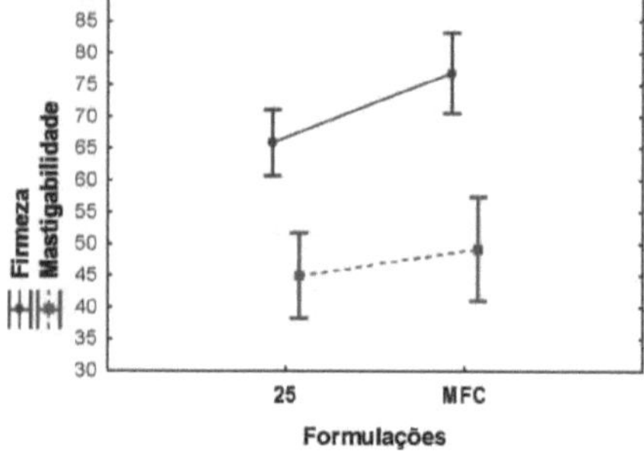

Figura 36   Firmness and chewiness graphs for comparison of chicken mortadella 25 and MFC. Source: the author.

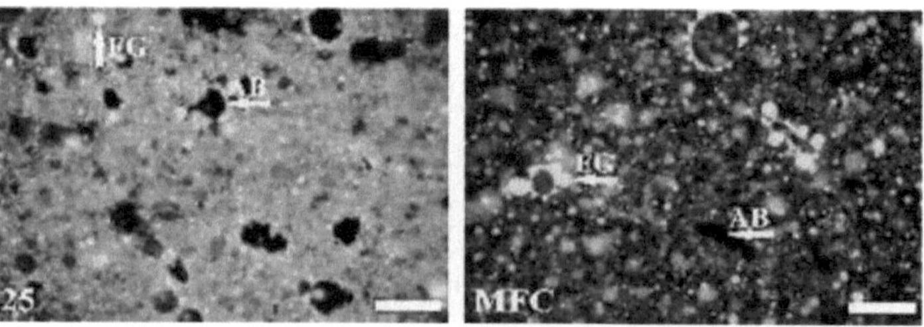

Figura 37   Fluorescence microscopy of chicken mortadellas. Formulations 25 and MFC. FG-Fat globules; AB-Air bubbles. Bar: 200jim.

Source: the author.

CHAPTER 5

CONCLUSION

The basic chemical components of emulsion formation (protein, fat and moisture) were assessed for their quality in the formulation of chicken mortadella. By relating them, the approach chosen for this study allowed the effects of these variables on the physical and chemical characteristics of the final product to be quantified and better understood. In general, many studies compare formulation results without taking composition into account.

The study showed that there are regions of different product stabilities that allow for greater or lesser loss during cooking, with a consequent change in the mechanical quality of the product. The optimum mixing range can be defined as (14-16% protein, 24.5-30.0% fat and 54.0-59.5% moisture). This region of greater stability produced chicken mortadellas with lower cooking losses and greater mechanical resistance, and the moisture/protein ratio showed a strong dependence between these variables related to cooking losses and the parameters firmness and chewiness.

In continuity, the addition of ingredients such as mechanically separated chicken meat (MSM), concentrated soya protein (CSP), cassava starch (MS) and sodium tripolyphosphate (STP) to the basic formulation of chicken mortadella had significant effects on cooking losses, pH and texture parameters of the product:

- it was observed that the addition of mechanically separated chicken meat (MSM) only had an influence at 10% and higher quantities (up to 40%) did not significantly alter the cooking losses and texture profile of chicken mortadellas, only increasing their pH value. In this sense, the results of the experiments showed that its use favoured a reduction in cooking losses, as well as making it possible to reduce formulation costs by replacing chicken breast and fat up to its legal limit of use (max. 40%);

- there was a tendency for cooking losses to decrease as the soya protein concentrate (PCS) content increased. Increasing the concentration of this type of protein (3 to 4 per cent) to replace the meat protein proved to be effective in reducing cooking losses and maintaining the firmness and chewiness of the product. The variability of the molecular weights of this protein, as well as its solubility characteristics in water and saline solution, may have favoured improving the stability of the product by acting in conjunction with the meat protein.

- cassava starch (FM) in concentrations between (1 and 4 per cent) considerably reduced cooking losses, slightly altering the pH and texture parameters of chicken mortadellas. It has been shown that adding this ingredient to cooked products causes the water to be strongly bound during cooking;

- the use of the stabiliser sodium tripolyphosphate (0.5%) reduced cooking losses, mainly due to the increase in pH, which shifts the pH of the isoelectric point of the proteins, as well as increasing the ionic potential of the medium, possibly increasing the extraction of myofibrillar proteins. Together with the salt (S), it improves the water binding of the final product. The performance of this additive was very positive, because even with the reduction in the concentration of salt (S), formulation 23, there was a reduction in cooking losses.

Formulation 25 was comparatively close to the reference, Seara brand commercial chicken mortadella (MFC), especially in terms of chewiness. With regard to firmness, it is likely that the 14% protein content in the MFC nutritional table explains its higher value compared to formulation 25 (12% protein). In this sense, the results obtained in this study could be even closer to the physicochemical results of the mortadella used as a reference. The results obtained favour the investigation of the effects of new ingredients, taking into account the search for lower cost, texture profile and emulsion stability.

REFERENCES

AOAC. **Official methods of AOAC International.** 17th. USA, 2000.

BARBUT, S. Convenience breaded poultry meat products - New developments.
Trends in Food Science & Technology, v. 26, n. 1, p. 14-20, 7// 2012. ISSN 09242244. Available at:
< http://www.sciencedirect.com/science/artide/pii/S0924224412000027 >.

BARRETO, A. C. D. S. **Effect of the addition of fibres as fat substitutes in Bologna sausage.** 2007.
Faculty of Food Engineering, State University of Campinas, Brazil.

BEJERHOLM, C.; AASLYNG, M. D. COOKING OF MEAT. In: JENSEN, W. K. (Ed.).
Encyclopedia of Meat Sciences. Oxford: Elsevier, 2004. p.343-349. ISBN 978-0-12-464970-5.

BOURNE, M. C. Chapter 1 - Texture, Viscosity, and Food. In: BOURNE, M. C. (Ed.). **Food Texture and Viscosity (Second Edition).** London: Academic Press, 2002a. p.1-32. ISBN 978-0-12-119062-0.

_________ . Chapter 4 - Principles of Objective Texture Measurement. In: BOURNE, M. C. (Ed.).
Food Texture and Viscosity (Second Edition). London: Academic Press, 2002b. p.107-188. ISBN 978-0-12-119062-0.

BRAZIL. **MAPA - Ministry of Agriculture and Supply. Normative Instruction No. 04, of 5th April 2000. Technical Regulation on the Identity and Quality of Mortadella.** 2000.

CALDERON, F. L.; SCHMITT, V.; BIBETTE, J. **Emulsion Science - Basic Principles.** 2 Ed. France: 2007.

DOMÉNECH-ASENSI, G. et al. Effect of the addition of tomato paste on the nutritional and sensory properties of mortadella. **Meat Science,** v. 93, n. 2, p. 213219, 2// 2013. ISSN 0309-1740.
Available at: < http://www.sciencedirect.com/science/article/pii/S03091740012003002 >.

FLORES, M. et al. Effect of a new emulsifier containing sodium stearoyl-2- lactylate and carrageenan on the functionality of meat emulsion systems. **Meat Science,** v. 76, n. 1, p. 9-18, 5// 2007. ISSN 0309-1740. Available at: < http://www.sciencedirect.com/science/article/pii/S0309174006003469 >.

GORDON, A.; BARBUT, S. Meat batters: effect of chemical modification on protein recovery and functionality. **Food Research International,** v. 30, n. 1, p. 511, 1// 1997. ISSN 0963-9969. Available at: < http://www.sciencedirect.com/science/article/pii/S0963996995000224 >.

HEDRICK, H. B. et al. **Principles of Meat Science.** 3rd ed. Dubuque, Iowa: Kendall/Hunt Publishing Co, 1994. 354.

HERRERO, A. M. et al. Tensile properties of cooked meat sausages and their correlation with texture profile analysis (TPA) parameters and physico-chemical characteristics. **Meat Science,** v. 80, n. 3, p. 690-696, 11// 2008. ISSN 0309-1740. Available at: < http://www.sciencedirect.com/science/artide/pii/S0309174008000740 >.

HONIKEL, K. O. CONVERSION OF MUSCLE TO MEAT | Cold and Heat Shortening. In: JENSEN, W. K. (Ed.). **Encyclopedia of Meat Sciences.** Oxford: Elsevier, 2004. p.318-323. ISBN 978-0-12-464970-5.

HORITA, C. N. et al. Physico-chemical and sensory properties of reduced-fat mortadella prepared with blends of calcium, magnesium and potassium chloride as partial substitutes for sodium chloride. **Meat Science,** v. 89, n. 4, p. 426-433, 12// 2011. ISSN 0309-1740. Available at: < http://www.sciencedirectcom/science/article/pii/S0309174011001872 >.

IGNÁCIO, A. K. F. **Reformulation of the lipid profile of emulsified meat products added with linseed oil and herbs and spices: evaluation of physicochemical and sensory characteristics.** 2011. Food Technology UNICAMP, Brazil.

JIMÉNEZ COLMENERO, F. CHEMISTRY AND PHYSICS OF COMMINUTED PRODUCTS | Non-Meat Proteins. In: JENSEN, W. K. (Ed.). **Encyclopedia of Meat Sciences.** Oxford: Elsevier, 2004. p.271-278. ISBN 978-0-12-464970-5.

JOLY, G.; ANDERSTEIN, B. Starches. In: TARTÉ, R. (Ed.). **Ingredients in Meat Products - Properties, Functionality and Aplications.** USA: Springer, 2009.

KEETON, J. T.; EDDY, S. CHEMICAL AND PHYSICAL CHARACTERISTICS OF MEAT | Chemical Composition. In: JENSEN, W. K. (Ed.). **Encyclopedia of Meat Sciences.** Oxford: Elsevier, 2004. p.210-218. ISBN 978-0-12-464970-5.

KNIPE, C. L. SAUSAGES, TYPES OF | Emulsion. In: JENSEN, W. K. (Ed.). **Encyclopedia of Meat Sciences.** Oxford: Elsevier, 2004. p.1216-1220. ISBN 9780-12-464970-5.

LAM, R. S. H.; NICKERSON, M. T. Food proteins: A review on their emulsifying properties using a structure-function approach. **Food Chemistry,** v. 141, n. 2, p. 975-984, 11/15/ 2013. ISSN 0308-8146. Available at: < http://www.sciencedirect.com/science/article/pii/S0308814613004846 >.

LEMOS, A. L. D. S. C.; HAGUIWARA, M. M. H.; YAMADA, E. A. Meat **Sausage Processing.** 2 Ed. Brazil: ITAL, Meat Technology Centre, 2011.

MANDIGO, R. W.; ESQUIVEL, O. CHEMISTRY AND PHYSICS OF COMMINUTED PRODUCTS | Emulsions and Batters. In: JENSEN, W. K. (Ed.). **Encyclopedia of Meat Sciences.** Oxford: Elsevier, 2004. p.266-271. ISBN 978-012-464970-5.

MCCLEMENTS, D. J. **Food emulsions: Principles, practice and techniques** 2nd ed. Boca Raton, FL, USA: : 2005.

MILLS, E. ADDITIVES | Functional. In: JENSEN, W. K. (Ed.). **Encyclopedia of Meat Sciences.**
Oxford: Elsevier, 2004. p.1-6. ISBN 978-0-12-464970-5.

MORIN, L. A.; TEMELLI, F.; MCMULLEN, L. Interactions between meat proteins and barley
(Hordeum spp.) p-glucan within a reduced-fat breakfast sausage system.
Meat Science, v. 68, n. 3, p. 419-430, 11// 2004. ISSN 0309-1740. Available at: <
http://www.sciencedirect.com/science/article/pii/S030917400400110X >.

NISHINARI, K. et al. Soy proteins: A review on composition, aggregation and emulsification. **Food
Hydrocolloids,** v. 39, n. 0, p. 301-318, 8// 2014. ISSN 0268- 005X. Available at: <
http://www.sciencedirect.com/science/article/pii/S0268005X14000319 >.

OCKERMAN, H. W.; BASU, L. CHEMISTRY AND PHYSICS OF COMMINUTED
PRODUCTS | Other Ingredients. In: JENSEN, W. K. (Ed.).
Encyclopedia of Meat Sciences. Oxford: Elsevier, 2004. p.283-290. ISBN 978-012-464970-5.

ORDÓNEZ, J. A. et al. **Food Technology - Animal Origin.** Madrid Spain: 2005.

PARDI, M. C. et al. **Meat Science, Hygiene and Technology.** Goiânia - Goiás - Brazil. II 2007.

PEARSON, A. M.; GILLETT, T. A. **Processed meats.** 3 Ed. New York: Chapman & Hall, 1996.

PINCUS, M. R. 2 - Physiological Structure and Function of Proteins. In: SPERELAKIS, N. (Ed.).
Cell Physiology Source Book (Third Edition). San Diego: Academic Press, 2001. p.19-42. ISBN 978-
0-12-656976-6.

PRICE, J. F.; SCHWEIGERT, B. S. **Science of Meat and Meat Products.** United States: 1971.

SEBRANEK, J. G. Basic Curing Ingredients. In: TARTÉ, R. (Ed.). **Ingredients in Meat Products -
Properties, Functionality and Aplications.** USA: Springer, 2009.

SUN, X. S. 9 - Thermal and Mechanical Properties of Soy Proteins. In: WOOL, R.
P. and SUN, X. S. (Ed.). **Bio-Based Polymers and Composites.** Burlington: Academic Press, 2005.
p.292-326. ISBN 978-0-12-763952-9.

TESTER, R. F.; KARKALAS, J.; QI, X. Starch-composition, fine structure and architecture. **Journal
of Cereal Science,** v. 39, n. 2, p. 151-165, 3// 2004. ISSN 0733-5210. Available at: <
http://www.sciencedirect.com/science/artide/pii/S0733521003001139 >.

TORNBERG, E. Effects of heat on meat proteins - Implications on structure and quality of meat
products. **Meat Science,** v. 70, n. 3, p. 493-508, 7// 2005. ISSN 0309-1740. Available at: <
http://www.sciencedirect.com/science/article/pii/S0309174005000434 >.

XIONG, Y. L. CHEMICAL AND PHYSICAL CHARACTERISTICS OF MEAT | Protein
Functionality. In: JENSEN, W. K. (Ed.). **Encyclopedia of Meat Sciences.**
Oxford: Elsevier, 2004. p.218-225. ISBN 978-0-12-464970-5.

ZORBA, O. The effects of the amount of emulsified oil on the emulsion stability and viscosity of
myofibrillar proteins. **Food Hydrocolloids,** v. 20, n. 5, p. 698-702, 7// 2006. ISSN 0268-005X.
Available at: <
http://www.sciencedirect.com/science/article/pii/S0268005X05001475 >.

ZORBA, O.; KURT, §. Optimisation of emulsion characteristics of beef, chicken and turkey meat
mixtures in model system using mixture design. **Meat Science,** v.
73, n. 4, p. 611-618, 8// 2006. ISSN 0309-1740. Available at: <
http://www.sciencedirect.com/science/article/pii/S0309174006000714 >.

Printed by Books on Demand GmbH, Norderstedt / Germany